SAVING THE GUINEA HOGS

The Recovery of an American Homestead Breed

Saving the Guinea Hogs

The Recovery of an American Homestead Breed

Cathy R. Payne

Rose Garden Press

© Cathy R. Payne 2019

All rights reserved. No part of this publication may be reproduced, distributed, or transmitted in any form or by any means, including photocopying, recording, or other electronic or mechanical methods, without the prior written permission of the publisher, except in the case of brief quotations embodied in reviews and certain other non-commercial uses permitted by copyright law.

Extensive quotes from *Managing Breeds for a Secure Future*, 2017, permitted by written authorization from authors D. Phillip Sponenberg, Alison Martin, and Jeannette Beranger

Rose Garden Press is an imprint of
Cedar Springs Garden Enterprises LLC
1860 Barnett Shoals Road
Ste 103-516
Athens, GA 30605-6821

Library of Congress Control Number 2019903124

ISBN 978-1-7335932-2-9 Ebook
ISBN 978-1-7335932-0-5 Softcover
ISBN 978-1-7335932-1-2 Hardcover

Photos by Gabi Rosenthal
Boar graphic, interior and cover design by Danijela Mijailovic

Printed by IngramSpark of Ingram Content Group LLC
1 Ingram Blvd, La Vergne, TN 37086

Praise for Saving the Guinea Hogs: The Recovery of an American Homestead Breed

"I thoroughly enjoyed reading *Saving the Guinea Hogs*. It was like sitting down with the breeders of old and listening to their guidance. Cathy R. Payne has written the definitive history of the Guinea Hog and told us how to take them into the future. I'm looking forward to the other books in this series."

Matthew Hunker, NC, Mechanic and Guinea Hog Breeder

"*Saving the Guinea Hogs* is an important work, especially because Cathy Payne is able to take the basic problems facing local and landrace breeds and is then able to show what sorts of practical steps can be taken to assure their survival as productive and adapted genetic resources"

D. Phillip Sponenberg, DVM, PhD, Professor, Pathology and Genetics, Virginia-Maryland College of Veterinary Medicine

"Dr. Payne's *Saving the Guinea Hogs* is more than a study of lineage and breeding. It is a powerful story about the role this special animal plays in peoples' lives. I could not decide which was more fascinating, the American Guinea Hogs, or their breeders!"

Kirstin L. Elsworth, PhD., Assistant Professor of Art History, California State University

"Cathy Payne has worked tirelessly to preserve both the genetics and the history of the Guinea Hog breed. She has taken a variety of information over several years and put each piece together to solve the puzzle of the mostly unpublished history to inform future breeders."

Lori Voigt, CART provider/transcriber for hearing impaired and deaf, ranch wife

"*Saving the Guinea Hogs* is a robust literary resource for those seeking to understand more about American homestead hog breeds in general and the Guinea Hog breed specifically. As I read the book I was educated about the American Guinea Hog. Along the way there were interesting diversions into southern agricultural history, socio-economic implications, threats of breed extinction, and the culinary characteristics of this special hog breed. The book also offers a heartfelt glimpse into American agricultural history and heritage through the personal voices of those interviewed."

Shane McGrath, corporate sales, musician, writer

Dedication

Dedicated to Gary Sumrall (1954-2016) and to *all* breeders who devoted time, labor, resources, and affection to their Guinea Hog herds from 1860 to 2018. Each breeder made a difference.

Table of Contents

Foreword by Dr. D. Phillip Sponenberg	xi
Author's Preface	xii
Organization of the Book and its Companion Website	xiv

Part I - An Introduction to the Hogs — xvii
Introduction	1
Chapter One Meet the Guinea Hog	7
What is a Guinea Hog?	7
History of the Name	16
Culinary Qualities of the Guinea Hog	17
Historical Records and DNA Evidence	21
Chapter Two Issues of Black and White (And Red or Blue all Over)	25
Color Variations in Guinea Hogs	25
Color Controversies	27
Historical Data and Traditions of Long-time Breeders	30

Part II - Raising Guinea Hogs 1940-1995 — 37
Chapter Three Era of the Southern Hog: Interviews with Elders	39
Fewer Fences	39
Butchering Time	41

Part III - Why did the Guinea Hogs approach Extinction? — 45
Chapter Four Vanishing Hogs	47
Changing Times	47
Consumer Demand and Industrial Production	50
Decline of the Family Farm	54
The Pet Pig Craze	55

Part IV - Organizing to Save the Guinea Hogs 1990-2006 — 57
Chapter Five Getting Organized	59
Guinea Hog Breed Organizations Past	59
Organizing the American Guinea Hog Association	62
Foundation Hogs of the AGHA in 2006	66
Table 1	67

Part V - Foundation Stock and Breeders in the AGHA	69
Chapter Six Bill Biggers	71
Chapter Seven J. Frank Baylis	75
Baylis VA Samson	75
J. Frank Baylis	79
Chapter Eight The Setty Line Hogs: Setty, Celesky, and Watkins Connection	83
Mark Celesky	83
Randy Setty's Herd	85
Setty Lilly	88
Setty Rose and Setty Houdini	88
Setty's April 2002 Litter	90
Celesky Hogs at Sedgewick County Zoo	92
Close Relations in the Setty Line	93
Part VI - Discovering the Missing Genetics	95
Chapter Nine Breeders of the Lost Genetics	97
Introduction	97
Billy Frank Brown and Fred Keene	98
Dan and Shirley Hale	104
Annette Hesters	106
Donna Watkins	108
Marcia Read	110
The Sumrall Family	116
Part VII - Expanding the National Herd	123
Chapter Ten Genetic Recovery Begins	125
My Story and my Process	125
An Email from Brent	127
Journey to Mississippi	128
Surprise from Indiana	131
Chapter Eleven The Historic Herds Network	133
Organization of the Historic Herds Network	133
More Hogs to Recover	136
Deborah Baker	138
Donna Dorminey	143
Becky Mahoney	146
Cathy R. Payne	149
Another Trip to Mississippi	152
Sumrall Feeder Hogs	157

Chapter Twelve The Genetic Recovery Project — 159
    Getting Started — 160
    But Wait, There's More! Donna Discovers the Maveric Herd — 161
    The Historic Herds Network — 165
    DNA Testing — 168
    The Genetic Recovery Process — 170
    Table 2 — 172

Part VIII - Where do we go from Here? — 173
Chapter Thirteen Keeping our Momentum: a Call to Action — 175
    Know Where you are Going — 175
    Breeding Strategies — 176
    Selection — 178
    Involvement — 181
    About the Author — 183
    Acknowledgements — 185
    Resources — 189
    Glossary — 191
    Appendix — 195
    Index — 205

Foreword

Saving the Guinea Hogs is an important work, especially because Cathy Payne is able to take the basic problems facing local and landrace breeds and is then able to show what sorts of practical steps can be taken to assure their survival as productive and adapted genetic resources. The issues surrounding Guinea Hogs are complex. They serve as a great model for saving other landraces and local breeds.

    The combination of tradition, wisdom, organizing skills, and education that were brought to bear on the challenges facing the breed all worked together for a constructive and fruitful outcome. The detective work was essential and difficult, and was followed up by wise and heroic actions that succeeded in salvaging rare and overlooked genetics in this unique and important American breed.

    Teamwork and inclusiveness are obvious at all steps along the way, and are especially worthy of attention because they serve as an example across all breeds: working together is the best way to save these breeds and to help all breeders!

D. Phillip Sponenberg, DVM, PhD,

> Honorary Member of the American College of Theriogenologists
> Professor, Pathology and Genetics
> Virginia-Maryland College of Veterinary Medicine
> Blacksburg, Virginia 249061 USA

Author's Preface

The concept for *Saving the Guinea Hogs* began taking shape back in 2013, when I was researching American Guinea Hogs. I looked for any and all information I could find about the hogs. Eventually I found a few helpful websites with snippets of information, magazine articles, and a chapter in a book here and there. There really was nothing in depth, and no books on the topic. All I wanted at first was a book telling me how to raise and care for the hogs. I decided that if I wanted to read that book, I had to write it first!

When I wrote down the list of everything I wanted to learn, I realized that there were several different topics. It would take years to learn about and write that much information. I decided I would divide up all the topics across at least three books. One book about the care of the hogs, one book about the meat and lard products it produces, and another about the genetic lines of the hogs. At the time, I thought I would write them in that order. That has now reversed, as I learned more than I ever believed I would.

Since the American Guinea Hog is an old-fashioned heritage breed that was almost extinct in 2006, I knew the best people to learn from were those who bred the hogs prior to that time and the ones who worked to preserve the breed and organize the American Guinea Hog Association (AGHA). My doctoral dissertation, completed in 2004, was an interview project that involved recording conversations, transcribing, analyzing, and writing a 236-page book. What better way to learn than to have conversations with people about the hogs and find out what it took to save the breed? I had the desire, organization, and writing skills to get the job done. I put those intentions out, and somehow attracted the right people to me. Over the last few years, I received just the information I needed at just the right time.

Because I had an original audience of one (me), this is a very personal book. In addition to citing sources and excerpting sections of my interviews, I have inserted my own opinions and conclusions. I admit to my own biases. I have attempted to be clear when I am the source of information. I am a lifelong learner, and a lifelong educator. If, by reading this book, you learn something about the Guinea Hogs, their history, genetic relationships between strains and with other swine breeds, their importance as genetic packages, and ways you can help assure their existence in the year 2120, then I have accomplished what I set out to do in writing the book. I wrote *Saving the Guinea Hogs* for breeders, homesteaders, farm-to-table chefs, sustainable farmers, hobby farmers, historians, conservationists, animal lovers, and anyone with a curious and open mind.

Organization of the Book and its Companion Website

I've separated the book into eight parts that are loosely organized by a timeline. There is a little jumping around in time, so key points may be repeated throughout. The Resources section in the back will point you to additional information and list my Bibliography and References. Use of the Glossary should help with acronyms and terms specific to pig farming and genetics, in case these are not familiar to you.

The Table of Contents and Index will help you to retrieve information quickly and I've provided an appendix to highlight a few historical documents. I did not include photographs in this book to keep costs down for the readers. However, I have a useful companion tool to the book at https://guineahogbooks.com/photo-gallery/. This is a gallery containing photographs, pedigrees, registrations, and descriptions of the hogs. They are divided by family strains so you are able to see the differences and similarities as you read, or whenever you have time to explore the site.

I plan to continue adding photographs and pedigrees to the gallery as often as breeders send them to me. Having these resources on a website allows the information to be updated regularly and provides important data on breeding, birth defects, or temperament changes that could not be included in a static print book. The gallery includes a search engine so you can search for specific hogs or specific farm prefixes. Eventually the site will include generations of hogs linked to each other. This will allow you to observe consistencies and differences within and between genetic lines. Owners of hogs provide photographs, descriptions, and pedigrees.

This visual and narrative site will help breeders decide whether they are interested in conserving a particular strain. They may choose linebreeding and use the site to locate farms with a different strain in order to linecross, for example. This fluid and dynamic website is more comprehensive than I could have included in a book. I hope that Guinea Hog Breeders will return and use this valuable tool to inform their breeding.

Part I
An Introduction to the Hogs

Introduction

> "Extinction is forever; you don't get them back.
> Our position is: Don't throw the genes away,
> your grandchildren might need them."
> *Hans Peter Jorgensen*

I get frequent emails from new Guinea Hog breeders wanting to know where they can find books about this breed. Up until now, they really couldn't. I saw the need for this information because I needed it myself. *Saving the Guinea Hogs* will tell readers what a Guinea Hog is—and what it is not. It will explain how the breed association for American Guinea Hogs came to be. I cover social, political, and cultural reasons that the hogs became nearly extinct. I share how they were obtained and bred, and which strains were not included in the gene pool of the registry for about ten years. You will learn how the hogs were raised historically in the South, how butchering was a family affair, and the emotional connections families had to their herds.

    I tell the story of a small group of breeders who located the missing strains and worked in conjunction with the American Guinea Hog Association (AGHA) to return the missing genetics. Ultimately, these genetics greatly enlarged the biological value of the breed. I present the information in a thorough but non-technical way so the average reader can readily understand some rather complex factors. In the final chapter, I try to make sense of what I've learned and make recommendations for breeders and others to support the maintenance and growth of this breed for the future. Along the way, you will meet some colorful characters from around the country, several southern gentlemen eager to talk about their childhood memories, and some endearing hogs with plenty of charm.

The Guinea Hog is a rare, heritage livestock breed. It is much smaller than those raised for commercial production, but still a productive working breed that is livestock; not a pet. There was a time when North America produced hundreds of varieties of apples, squash, and tomatoes. Every family or community had a variety that they passed down through saved seeds year after year. Likewise, there was a wide variety of livestock breeds, each adapted to a local community.

Livestock breeds around the world are becoming extinct at a quick pace. The Livestock Conservancy monitors and assists in the conservation of about 180 rare livestock and poultry breeds in North America. Why is conserving them so important? The authors of *Managing Breeds for a Secure Future* (shortened to *Managing Breeds* in this book) write, "Breeds serve as the main reservoirs of the genetic diversity within a species. Half of the biodiversity of most domesticated species is shared across breeds, while the other half is unshared and is instead contained only within single breeds. The consequence of this is that losing breeds means losing genetic diversity, because by losing a breed the species loses the genetic information that is unique to that breed."

They also state that, "Entire breeds, rather than just the individual component genes of individual animals, need to be conserved as intact genetic packages for their potential use in agricultural systems or in other services to humans."

In Kathleen Maclay's article "Breeding Rare Animals Helps to Preserve Diverse Gene Pools" featured in the *Los Angeles Times*, she

INTRODUCTION

speaks with Hans Peter Jorgensen. Jorgensen was the manager of the C.S. Foundation Farm. He warned, "Extinction is forever. You don't get them back…Don't throw the genes away, your grandchildren might need them."

The Guinea Hogs' DNA has been compared to that of several other heritage swine breeds around the world. The Guinea Hog tends to be an outlier that is not closely related to many of the breeds tested. Therefore, it is especially important to protect this unique genetic information. The hog, like its DNA, is unique as well, and especially suited for homesteads, farmsteads, and small landholders. This Southern hog, like the antebellum homes in the town where I live, is part of our history and culture and worth preserving. As homesteading, sustainable farming, and farm-to-table restaurants become more common and popular, these hogs will have an opportunity to prove their value.

When I discovered that there were no books written about the American Guinea Hog, it seemed natural that I would learn as much as I could about my own herd and locate people who could tell me more about the hogs' history. I tracked down every lead and asked question after question. One interview led to another. For my process, I used a digital recorder for live interviews and the service Recordia Pro to record phone conversations. I then converted each recording to an MP3 file. I had each interview transcribed so the transcribed text became my database. I scanned each transcript for information, dialogue, and categories. The categories helped to determine the sections of this book.

Like climate change, rare breeds are at a tipping point that can go either way. What we do at this juncture in time is important. Decisions are made by individual breeders with the needs of the specific breed in mind. Each individual animal holds unique genetic information, and the overall herd holds unique information. Once lost, it cannot be recovered.

Breed associations and The Livestock Conservancy can make recommendations to breeders about concerns for the overall herd or flock in question. If numbers are low, numbers must be increased. If certain strains or family lines are missing or low in numbers, those specific numbers must be increased. Possible genetic abnormalities must be eliminated. Breeds need promotion to those who may want to breed them. Breed products such as wool, meat, or dairy must be promoted to potential customers and chefs. Breeders and consumers need to define their goals and learn the strategies that will give them the outcomes they want for future generations. My hope is that this book may inspire you to take some small or large action in one or more of these directions. It may be that you visit a heritage farm, buy meat from a heritage cow or hog breed, or purchase a scarf made from heritage wool. It is important to "Eat 'em to save 'em" and to "Shave 'em to save 'em." If a breed does not have a useful product to sell, there is no incentive to continue a breeding program.

Putting this story together was a bit like putting together a jigsaw puzzle. I started with a lot of unrelated puzzle pieces in the form of interview transcripts. After getting twenty or so interviews under my belt, I started to see connections, patterns, and similarities. Continuing the puzzle analogy, I put the flat-sided pieces together to make an edge (all the interviews that mentioned the same person, breeder, or event). Eventually, I filled in the corners and a picture began to form. I found themes—hogs in the woods, hog butchering time, hogs becoming scarce, etc.

At the beginning of my research, I learned that there were several family lines of the hogs that existed but were not included in the only formal registry for Guinea Hogs with the AGHA. During my interview process, from the end of 2014 to early 2016, I worked with a team of breeders to recover those family lines. We eventually succeeded in infusing the AGHA registry with the genetic diversity

INTRODUCTION

needed to keep the national breed healthy and diverse for future generations.

Because I was actively involved in the process, this section of my book will include characteristics of memoir—a story I remember, a focus on the relationship between particular people and animals and me, and explanations of the significance of these relationships and activities that are focused on a particular period of time. You can expect me to pop in and out of the story throughout this book, inserting my opinions and observations.

The research process has been intriguing, enlightening, humbling, and very rewarding. I am honored to have met so many colorful, hardworking people. I was privileged to listen to these stories and preserve them for future generations. The journey included many twists, turns, and surprises. My story begins with a primer on the Guinea Hog.

Chapter One
Meet the Guinea Hog

"These were the best hogs I've ever seen."
Cohen Archer

"When one variety suffers destruction, entire plant and animal populations can be irretrievably lost; uniformity engenders vulnerability."
Hans Peter Jorgensen

What is a Guinea Hog?

I like to refer to the Guinea Hog breed as "hogs with heart." It is easy to bond with these gentle animals: so full of personality and intelligence. Hogs are, at their core, social creatures. Historically, this hog has meant a lot to Southerners. The hogs provided meat, lard, and sometimes income for the family. Hogs in general are called "mortgage lifters." The resources of the hog helped cash-strapped farm families find a way to make their mortgage payments. In addition, the old breeders hold fond memories and emotional connections to the hogs. Those who raised a variety of hog breeds made a point of telling me that they strongly favored the Guinea Hog breed.

In the early spring of 2017, I was raising Guinea Hogs in Northeast Georgia. I had spent a couple years seeking out older Guinea Hog breeders so I could pick their brains about the history of the hogs. It was getting increasingly difficult to find first person documentation.

One morning in March, I was a little late getting out to tend the livestock. I was surprised when the house phone rang, as it was only eight o'clock.

"Good morning, this is Cathy," I answered.

The raspy voice of an older southern gentleman replied, "My phone told me to call you." I didn't recall leaving any messages, but I bit my tongue and listened.

"My name is Cohen Archer. Do you have any Guinea Hogs?"

"Yes, I do," I said.

"Do you have the big-boned or the little-boned?" he inquired.

"Well, I have a nice mix of both, but I think mostly big-boned," I replied.

As we made arrangements for a visit, Mr. Archer told me that he was seventy-five years old and from Washington County, Georgia. He said he had not seen a Guinea Hog in a long time. He started telling me about the hogs his daddy kept. "They gained weight easily," he told me. "My family ran them in the woods. They had their babies in pine sapling pasture."

I told Mr. Archer that I was saving stories like his and would be honored to have him visit the hogs as soon as possible. I requested and received permission to tape record his stories to share with others. Before we hung up, I asked Cohen how he had found me.

"I told my phone to find Guinea Hogs in Georgia or South Carolina," he answered, "and your name and phone number showed up on my phone." Siri to the rescue!

On the appointed day, Cohen and I walked the pastures so I could show him my beloved breeding sows and boars. I turned on the recorder as I took him to see a brand-new litter of piglets that my sow, Yokeley's Summer Thyme, had delivered.

"That is about what we used to have when we used to grow them, when I was a boy," Mr. Archer reminisced. "That is a typical Guinea Hog, as I remember. They were very gentle. I used to take care of the pigs. If you have them out in the open, [the sows will] make their own bed. It will be a huge pile of straw, and different things like sticks that they pick up. [The sows] put the pigs up under the straw and all. And you won't even see them until [the sow] calls them out. These were the best hogs I've ever seen. We just had them running in the woods, and we would fence off the woods. We grew crops in the upper part of the land, and we would fence them in when we started planting crops. And every day we would call them up and give them a little feed to keep them coming. Just really nice, gentle hogs. We named some of the sows.

"We had more breeds than just the Guinea Hogs, but we liked the Guineas better. They would just make it on their own. Some of the hogs would lay on the pigs, but with the Guineas, we didn't have that problem."

"Did you know anyone else who had Guinea Hogs?" I inquired.

"I didn't. Didn't really know how important it was to keep in touch with them. But now that I am getting old, I see. They say your hindsight is better than your foresight."

I took him over to see a young boar that had flopped over to be rubbed on his belly. Cohen bent over and started rubbing the soft, warm hog. It began to grunt appreciatively. Another boar came over and offered himself for a rubdown too. Cohen had one hand on each hog.

"Isn't that a nice hog?" he exclaimed. "That is just nice, I'll tell you!" Cohen seemed suddenly transformed in time from an elderly gentleman to a young boy again. His voice was excited and awestruck. "I remember those curly tails. As a child, I loved the Guinea piglets when they were young. They were so pretty and plump. I always remember that. And I remember when they were expecting; they would be huge! I wish I could remember all about them. I've been thinking about my daddy's hogs for a long time. Just last week I said, 'Well, I'll just see if I can find some. And that is what I done.'"

I did a little math in my head. The Archer family had raised hogs for nine years, from 1946 to 1954. Cohen was now seventy-five years old. His father had died when he was about twelve years old. "So, it has been sixty-two years since you have seen or touched a Guinea Hog!" I exclaimed. "You called me out of the blue because you wanted to see one again. And you didn't even know I was here until you asked your phone to find me."

"Right." I think at that point we were both choked up. We each wiped back tears.

"How do you feel, reliving that childhood memory?" I asked quietly. My throat felt thick.

"It feels really good," he said with a sigh.

Cohen Archer could never forget the Guinea Hogs, but his children and grandchildren never got to meet one. He hoped to get his son over to visit, and a nephew, but it never happened. By preserving the memories of the elders I've spoken with, I hope to help readers who have never met the hogs to understand their role in American history. For those readers who do breed the hogs, I want you to know and understand the significance of their history and the important role these hogs hold as genetic resources. It is my hope that your children, grandchildren, and great-grandchildren have at least the opportunity to meet a Guinea Hog one day.

Guinea Hogs are an even-tempered homestead hog, suited for small landholders. According to oral records, they were once common in the Southeast. Guinea Hogs are a small, dark-skinned local hog. They have erect ears and a single curl in their tail. Their bristles are typically black and can be long or short, thin or thick, and either straight, wavy, or curly. Hogs with the heaviest dense coats, especially curly types, may shed heavily during the summer. Noses vary from short to medium or long. Eyes usually face forward and have an alert, intelligent expression. The eye color might be dark brown, bright brown, or gray (blue). Their darkly pigmented skin and coat protect them from sunburn during the intense sun of hot southern summers.

Although the outward phenotype can vary widely, the breed overall is friendly, gentle, and easy to manage. Sows often allow trusted caretakers to handle piglets without objection. Homesteaders with children, farmers new to swine, and middle-aged farmers especially appreciate this calm temperament.

Like other swine, Guinea Hogs can be trained to respect electric fences, follow routines, come when called by name, sit on command, and more. They will use their shovel-like noses to dig up pastures, especially after a rainfall, but do far less damage than larger breeds.

Guinea Hogs sleep less and forage more than many other hog breeds. They spend their time gathering pasture grasses, roots, and acorns. Like all swine, they are omnivores. They may hunt for grubs, field mice, or snakes. They thrive on a quarter of the commercial feed required by modern "improved" breeds. Some bloodlines are a bit larger and grow quicker, while others are smaller and slower to finish. The slow growth, strong muscles, and varied diet all serve to enhance the flavor of the Guinea Hog's meat. They all taste delicious.

The Guinea Hog breed is unique to the continental United States as a local breed. There have been and still may be breeders in Alaska. Guinea Hogs are about half the size of typical hogs and grow more slowly than the commercial type. The Guinea Hog typically reaches 125 to 250 pounds on the hoof in twelve months and will yield 71 to 144 pounds of hanging weight meat. In contrast, other heritage breeds and commercial breeds can weigh 250 pounds on the hoof in half that time and yield 144 pounds of hanging weight meat.

It takes at least twice as long to grow the same amount of meat from a Guinea Hog. For this reason, meat growers cannot earn as much money selling pork from Guinea Hogs if they market and price it similarly to the larger hogs. However, there is a niche market at farm-to-table restaurants for their delicious slow-growing meat. On a small scale, Guinea Hog pork can command a premium price, and chefs appreciate being able to keep a whole hog in their coolers and cure the hams in just six months. Larger hams take up to a year or longer to cure. Freezer space is at a premium for many customers and homesteaders and some would rather not have so much meat to store.

Watching a herd of Guinea Hogs grazing on lush grass pasture is fascinating and hypnotic. They will snatch up grass as enthusiastically as a cow or sheep and actively engage in the collection of their own food. They will stay active most of the time when weather is cool.

In the heat of the summer, they will spend much of their time resting in the shade or soaking in mud wallows. In autumn, as grasses become dormant, the hogs are more than happy to munch on acorns, glean corncobs, or pick up windfall fruit in the orchard. Guinea Hogs are excellent foragers and leave nothing to waste. Breeders in northern states and the Midwest report that Guinea Hogs are quite hardy in winter climates. In every climate, they still require shelter at all times to protect them from wind, rain, snow, and sun. The hogs may choose to stay outside of the shelter, but the option for protection must be provided.

Guinea Hogs are reported to have been a well-established breed of pig in the South prior to the Civil War. After Reconstruction, many wealthier landowners switched to "improved" breeds of pigs. These standardized breeds, of Chinese and European origin, grew faster and brought a better price at market than the small, local, black hogs of the South. I've been told that Gullah populations on the South Carolina Coast, descendants of slaves, raised Guinea Hogs for a long time. Middleton Place Plantation in Charleston maintains Guinea Hogs in their barnyard because they found notes indicating they were historically raised there, as well. Guineas remained with people of modest means, both black and white Southerners, stuck with the Guineas because they could not afford to switch to a more expensive breed. This gave Guinea Hogs a reputation as a "poor man's hog." The Guinea Hogs were easier on small landholdings, cost little to rear, and were easy to handle.

Families that raised the hogs used the entire animal from head to tail. The lard was used for cooking, lubrication, and making soap and candles. The head was transformed into headcheese, or *souse,* and all parts went into scrapple. Hams and bellies were smoked or salted to preserve them for the winter. Intestines became chitterlings, and skins were made into pork rinds. The meat of the old-fashioned Guinea Hogs raised on pasture has a unique flavor, aroma, and texture. It is so delicious that the breed was added to Slow Foods' *Ark of Taste,* found on their online catalog. This is a listing of breeds worth preserving for their culinary qualities. I will delve further into the culinary aspect of Guinea Hogs in another section of this chapter.

The Guinea Hog is a landrace breed. I rely on *Managing Breeds*, by Sponenberg, Martin, and Beranger, 2017, to describe and explain the term, *landrace*. Quotes and paraphrasing to follow come from this book, and attributions to Sponenberg will include Martin and Beranger, as well.

Breeds are groups of animals with consistent, recognizable types. When mated together, they produce more of that consistent type. These readily identifiable genetic packages can be repeated and are predictable. Predictability is important for agricultural use because the farmer can expect consistent results. Sponenberg and coauthors discuss five classes of breeds. All five classes produce a consistent type, but the landrace and feral classes are more variable than the other three.

Sponenberg and coauthors state, "Landrace breeds represent an early stage of breed development. 'Landrace,' as used here, is a general term that refers to populations of animals that are isolated to a local area where local production goals and the local physical environment drive selection. The 'landrace' designation should not be confused with the specific Landrace swine breed, or with the Finnish Landrace sheep breed (now known as Finnsheep). The landrace concept, as used here, is important as a general pattern for many breeds of all species. Landraces are sometimes called local breeds, natural breeds, or primitive breeds."

The outward appearance of the animal is the phenotype, and is reflective of the unseen genotype, or underlying genetics. Sponenberg asserts that, "Entire breeds, rather than just the individual component genes of individual animals, need to be conserved as intact genetic packages for their potential use in agricultural systems."

When most of us think of breeds, such as with dogs, we think of a consistent appearance, size, and personality. Dog breeds are widely standardized. Dogs that meet certain standards are selected to breed with other dogs that meet those same standards to continue those genetic traits. Most swine breeds are also standardized. Some have strict breed standards and herd books that date back to the 1800s in Europe.

In contrast, there are a few remaining landrace breeds of dogs, swine, sheep, horses, and cattle. Many of those that do remain have never had a herd book or did not have one prior to the twenty-first century. Devoted conservationists initiated documentation of the landrace breeds in order to preserve them. Landraces are useful genetic resources because of their ability to survive with little inputs. Their history of resistance to diseases without intervention from man make them essential in the event that a disaster strikes among modern commercial strains.

The history surrounding these local breeds is likely to be in the form of oral history, rather than formal written documentation. Sponenberg explains, "Landraces are by their very nature more variable than are other classes of breeds. Landraces are characterized by a consistency of biological and adaptation traits, and not necessarily by uniformity of physical appearance. This superficial variability can lead many observers to dismiss landraces as trivial and unimportant when the reality is just the opposite. They are historically and biologically important genetic resources that are adapted to difficult environments. They are generally productive with few inputs, having excelled in survival and adaptation, although generally with less emphasis on high individual levels of production than is typical of the other classes of breeds."

Guinea Hogs share repeatable, predictable characteristics that make them an identifiable breed. However, as a landrace, they include variations, or strains. This is because when people were more isolated, different breeders selected the variations they liked or that survived that environment from the variations born in that place. This varied from breeder to breeder. There was no breed association, herd book, or standard of perfection. The current association for the breed, the American Guinea Hog Association (AGHA), was not formed until 2006. The association presents the breed standard in the form of a breed description, due to the variability of the landrace. The distinction of description versus prescriptive standard was specified in the original bylaws of the AGHA: "The Board shall have the authority to adopt and amend the "standard of type" which shall be presented to the membership in the form of an AGHA Breed Description." They did not define a hypothetical standard that would drive uniformity into the breed. Instead, the descriptive breed standard was written to define common traits and ranges that existed historically. This approach is recommended for landrace breeds because it not only helps retain the breed's traditional look and behavior, but also assures retention of the valuable genetic diversity that is present within the various strains.

Guinea Hogs are known to have a variety of body shapes and sizes. The larger variety was known as "big-boned," denoting a heavier bone structure. This size was reported to reach up to 450 pounds at maturity, according to interviews and descriptions. The bloodlines of the largest hogs died out by the 1970s, according to those same sources. An adult Guinea Hog weight of a big-boned type at four years is now more likely to max out at 250 to 350 pounds.

Setty bloodlines that include the boar Little Old Stiff Guy and Sumrall lines descended from the sow Sumrall Bobbie Sue tend to produce the big-boned type. The smaller variety was known as "little-boned," due to a smaller bone structure. The little-boned hogs can be as small as 120 pounds at maturity for a sow and 150 pounds for a boar. The Hesters bloodline of hogs is an example of the little-boned type.

James Priest, who raised Guinea Hogs as a young boy, was born in Kennesaw, Georgia in 1934. He told me a bit about the different body types. "We had one big-boned sow. It didn't work out with her. She would never have over five or six pigs. She was just so fat, and she was bad about killing her pigs. So we got rid of her, and we just kept the little-boned ones. We wound up with all little-boned. The big-boned was just bigger. I'd say they might weigh a hundred pounds more than the little-boned. But you know, raising pigs, you don't feed one mama hog for five or six pigs a litter. You want nine, ten, or twelve, something like that. We usually got nine or ten. And we kept them eight or ten weeks. But we always sold out.

"They were all solid black. They were black all over. I would say if you had one with some white on it, that it would have a mix in it other than a purebred, but it could be a throwback or something. What they used to call a throwback to a generation where they were breeding them up. I think ours were all purebred, but we never had papers or anything on them."

The American Minor Breeds Conservancy (AMBC), dedicated to preservation of American heritage livestock and poultry, retained an archival letter they received from Robert Gear in January 1989. He theorized that the Guinea Hog had three sizes: large, medium, and miniature. At the time, letters in local newspapers described Guinea Hogs that were mature at fifty pounds, along with the little- and big-boned sizes. This time period was also the beginning of a pet pig craze.

Miniature pigs were popular, and apparently Guinea Hogs and Vietnamese Pot-bellied Pigs were used in developing them. The Guinea Hog Association (GHA) recordation service was formed in 1991, and they included miniature sized Guinea Hogs in their breed description. Also during this period, the Sinclair Hogs and Minnesota Miniature strains were developed for use in medical research. Each of these mixed strains included Guinea Hog genetics in their development.

History of the Name

The Guinea Hog has been known in the southeastern United States by various names over the years. These include Guinea Forest Hog, Forest Guinea Hog, Pineywoods Guinea, African Guinea Hog, Miniature African Pig, African Pygmy, Guinea, African Miniature, Acorn Eater, Yard Pig, and, since 2006, American Guinea Hog. The term *Guinea* may have been used to indicate a smaller size at one point in our history. For instance, Guinea is a name sometimes used for a strain of

Criollo cattle known as Florida Cracker. They are smaller than most cattle—no more than 500 pounds.

The first person to suggest the name "American Guinea Hog" was Kirk Fackrell. In 2005, he purchased a gilt, Skyfire Heirloom, from Paul Krumm of Skyfire Farm. He also purchased a boar, Maveric Balthazar, from Arie McFarlen of Maveric Heritage Ranch. Kirk wanted the hogs shipped to his farm, Cascade Meadows, in Oregon. The Oregon Department of Agriculture refused to let the hogs enter the state because the name "Forest Guinea Hog" implied an African origin. Kirk changed the paperwork to call them American Guinea Hogs and added the Latin species name with it, he told me. After that, the state allowed the hogs to enter.

Kirk was active as a volunteer with the newly formed AGHA when it formed in 2006. He told the founders about his difficulty communicating to agriculture departments regarding the breed's domestic nature. He suggested changing the name to American Guinea Hog to help prevent similar problems in the future. Now, current breeders most often identify the breed as American Guinea Hog (AGH). However, the "old-timers" I interviewed for this history use the terms Guinea Hog, Guinea, or Guinea Forest Hog to talk about them. The Livestock Conservancy began researching the Guinea Hog in 1987 and that organization continues to call them Guinea Hogs. In this book, I will also use the traditional Guinea Hog designation, but may also refer to them as American Guinea Hog, Guinea, or AGH as appropriate.

Culinary Qualities of the Guinea Hog

Guinea Hog pork is not "the other white meat." This breed produces a juicy, flavorful pork with an incomparable mouthfeel. Its taste is influenced in part by the pasture grasses, dirt, vegetable scraps, acorns, legume hay, and windfall fruit on which it forages. These carbohydrate sources are converted to fat in the hogs and it is the fat that imparts the delicious flavors and textures into the Guinea Hog meat. Breeders, consumers, and chefs report a difference in taste when Guinea Hog meat is compared to breeds of hogs—even those also raised on the same land and fed the same diet.

Tammy Albert of Jail Creek Farm in Georgia told me her family conducted a taste test between Hampshire and Guinea Hogs, both pastured and raised on her farm. The Hampshire hog yielded a larger carcass. The difference between the Hampshire and Guinea Hog pork chops was the most noticeable, Tammy told me. The Hampshire meat was white and dense with a rubbery consistency on the fat. The Guinea Hog meat was darker in color, and the fat melted into the pan and into the mouth with a buttery, rich flavor. She reported "Flavor wise, once you taste AGH, it is hard to go back. A big difference in taste there."

Marcos Fernandez, chef-owner of restaurant Nineteen61 in Lakeland, Florida, told me about the first time he tasted Guinea Hog meat. "I was very impressed. I couldn't believe how different it was, and I enjoyed the meat more. My patrons really enjoy the meat, too. It is very sweet and different from a regular hog."

Around 2006, Arie McFarlen of Maveric Heritage Ranch nominated the Guinea Hog into the Slow Food, USA's *Ark of Taste* online catalog. Anyone can nominate a food for this designation, as long as it is from an endangered breed, prized by chefs and consumers, free of genetic modification, and fair (not commercial or trademarked). Local chapters of Slow Food USA promote connections that help people have equitable access to food. At their events, they aim to serve meals sourced from the *Ark of Taste* whenever possible.

In February 2010, Carlo Petrini, the Italian founder of Slow Food International, traveled from Italy to Georgia to attend a fundraising event at Watershed Restaurant in Atlanta. The goal was to raise funds for Petrini's annual international conference in Italy known as Terra Madre. He was already coming across an ocean to the Southeast for the Atlanta fundraiser, so Charleston Slow Food Chapter Leader, Carole Addlestone, decided to take advantage of his proximity.

She arranged for Petrini to travel from Atlanta to Charleston for an additional fundraiser with her chapter. It was a dinner event at Cypress Restaurant with Chef Craig Deihl. Charleston Slow Food wanted to highlight Guinea Hog on the menu because they were working with the American Livestock Breeds Conservancy (ALBC) to get more people raising the hogs. Jeannette Beranger worked for the ALBC and was friends with Gra' Moore, owner of Carolina Heritage Farm. Gra' (pronounced "Gray" and short for Graeme) began raising Guinea Hogs as a hobby in 2005.

Gra' explained, "Jeannette [Beranger] called me one day. I think I just had one boar and two sows back then. She told me she was going to go to Charleston to meet up with Carlo Petrini from Italy. They were looking for a rare breed to feature at one of the Charleston restaurants, and they wanted to know if I would be interested in donating a couple of Guinea Hogs. I kind of thought of doing this for a living but didn't know how to get my foot in the door. Craig Deihl at the restaurant was known for working with whole hogs, so I took a couple of them to him, and he really liked them. And it went from there."

Carole shared the story from her perspective. "Slow Foods USA had contacted me and said they wanted to put the Guinea Hog on the revival list, because you have to eat it to save it. They wanted to highlight that issue. They asked me to contact Gra' Moore who was raising a few of them at the time and see if he could raise some more. They wanted me to introduce Gra' to different chefs in Charleston to promote it, and that is what we did."

She continued, "I talked to Craig Deihl because I knew that charcuterie was his main deal, and he was always breaking down whole hogs and everything. I put him in touch with Gra' Moore and through that connection he got to Chef Sean Brock of Heritage restaurant (Also author of *Heritage*, a cookbook in which he features the Guinea Hog and other heritage breeds). Sean was interested and served it in his restaurant. So we kind of did it that way. It was very exciting working with the American Livestock Breeds Conservancy to bring it back. Carlo Petrini loved Charleston. I surprised everybody by having Ann and the Magnolia Singers do some plantation singing. And Gra' Moore gave a talk about the Guinea Hog before dinner." (Ann Caldwell and the Magnolia Singers perform Gullah poetry, stories and songs. Gullah are descendants of enslaved Africans who live in the low country of South Carolina and speak Creole English, or Geechee).

Chef Craig Deihl, while experienced with breaking down whole hogs, had not prepared Guinea Hog prior to the 2010 event. He shared his perspective on the Slow Foods Charleston dinner with me. "Slow Foods Charleston approached me about working with the American Guinea that Gra' Moore was raising in South Carolina and giving them my opinion on it. Not only for them, but for the American Livestock Breeds Conservancy. So it was one working with the other hand in hand. My responses were based on fat content, taste, tenderness, workability and use of the carcass—the whole nine yards.

"We cooked up the ham and tasted it. Was it tender, tough, meaty? Evaluating it on those regards. And it really is one of those strange oddity type of pigs, because it is a very small pig. The first hog was probably seven to eight months old and weighed eighty-eight pounds. That was the hanging carcass weight, so it was probably 115 or 120 [pounds] live weight. I cut it and tasted it. It was super flavorful. It was nice and tender. It kept on coming down to the abundance of fat. Excessive. And even at eighty-eight pounds, it was a pig that was excessively fatty."

He told me how he experimented with various ages and sizes of hog, presumably on later encounters. "So then we went down to fifty-five pounds, a five month old, and the meat was no longer rich and red. It was more pink in color, and the fat content was more of where it needed to be. So age was a factor in the color, fat, and flavor development. We played around with it. And I found that for cooking purposes the pasture-raised Guinea Hogs at six months old were perfect for cooking. The loins still had heavy intramuscular marbling in them. We can do pork three ways. We can do a piece of loin, a piece of chop, a piece of pork belly, and get eighteen orders out of a pig on the plate.

"We still make money on it selling that way. And then we have the hams and shoulders to do things with. We might do a sausage and wrap the shoulder around it, take the hind quarters, brine it, and do a fresh ham, or smoke it and do ham and eggs with collard greens. Even delicious little head cheese and egg dishes. With a larger hog we turn it into lard, pancetta, and other things. At that age the meat has a vibrant red color that looks more like beef when we trim it. We use a lot of lard. We make biscuits with it, fry potato chips or chicken in it, or confit chicken or duck."

Gary Sumrall, a third-generation breeder in Mississippi, shared a story about why the Guinea Hog was special to his family for 100 years. "We had the lard, plus the meat. The other hogs were pretty lean in comparison. My grandmother, my mother's mother, she'd come and help kill hogs on Thanksgiving each year. And when she died, she was ninety-three. And her twin sister, when we had a visitation, she said, 'Poor Willy, that meat finally killed her. ' (Laughter) We had cracklings out of it, and we'd eat a lot of crackling corn bread. Daddy had a smoke house and he smoked his own bacon and his hams. After he smoked 'em a couple of weeks, we started in on 'em. We would sneak to the smokehouse and get us all some. We got a lot of whippings over that. But it was good. It was hickory wood smoked, and it was good. But I do remember that well."

For the reader who is getting a bit hungry right now, and has never tasted Guinea Hog, I recommend that you remedy that situation just as soon as possible.

Historical Records and DNA Evidence: The Improved Essex Connection

The origin of Guinea Hogs is still being explored. Stories have evolved over the years based on historical and DNA data. Guinea Hogs were on the radar of the AMBC as early as 1987. Here is an excerpt from that organization's *Minor Breeds Notebook* published in 1989:

"Guinea Hogs, or Guinea Forest Hogs, most likely originated in the Guinea coast of Africa and were spread widely through the slave trade from Africa to England, France, Spain, and America. At one time they were common homestead pigs in the southern U.S., but are now practically unknown. Guinea Hogs were also used for breeding with English pigs in the 1700 and 1800's, and the very distant relationship between the two types made for an excellent cross."

In 1991, Donald E. Bixby, DVM and executive director of the AMBC wrote an article for the Guinea Hog Association's spring newsletter entitled, "Pigs as Pets." He wrote, "The Guinea Hog came from Africa to the South in the slave trade and has provided food for the rural poor in the U.S. for generations. The breed is documented in historical journals as a huge, red, hairy pig. What we call Guinea Hogs today are small and black." Stories about Guinea Hogs coming from Africa with the slaves were common before 2010, but now DNA evidence seems to be telling us a different story.

I've heard reports that Guinea Hogs were listed as inventory in Thomas Jefferson's Monticello. However, Jefferson's hogs were solid red—not black—and had a large build, not a small one. It is currently believed that the modern Guinea Hogs and historically red Guinea Hogs were the same in name only, and not genetically linked.

The Livestock Conservancy received a Sustainable Agriculture Research and Education (SARE) grant to explore DNA of several swine breeds. The rationale was that rare breeds of pigs are a "genetic reservoir for regional adaptations, biological fitness, maternal skills, foraging ability, lard production, and disease resistance." The study, which started in 2008 and concluded in 2015, had five objectives. The first objective was to quantify the genetic variability and the genetic relationships within and among rare swine breeds using DNA and pedigree analysis.

The DNA analysis was completed in Canada and involved collecting blood and hair samples from nine endangered breeds of pigs. These included Choctaw, American Guinea Hog, Gloucestershire Old Spots (GOS), Hereford, Large Black, Mulefoot, Ossabaw Island, Red

Wattle, and Tamworth. Data were analyzed using multiple methods. In each instance, Guinea Hogs clustered with Gloucestershire Old Spots, indicating that they are genetically similar. Both the GOS and Improved Essex (thought to be an ancestor of the Guinea Hog) have unimproved Berkshire in their foundations, and GOS and AGH have some common foundation. This is evidence that the Guinea Hog could be at least partially derived from the Improved Essex hog, a small black breed considered extinct in 1967. In conclusion, the DNA analysis indicates a European connection, not an African one.

The Essex was introduced to the United States in the 1800s and became very popular until after the Civil War, but was gone by the mid-1900s. That timeframe makes it possible to have contributed to the development of the Guinea Hog before the Civil War. The Essex hog may have also been used to develop the Small Black or Black Suffolk breed that died out at the end of the twentieth century. That hog also has much in common with today's survivor, the Guinea Hog.

The improved Essex breed was originally recorded in 1896 in *Record of Improved Essex Swine, Volume III*, as "Color black with no white whatsoever; face short and dishing; ears small, soft, and standing erect while young but drooping slightly with increased age; carcass long, broad, straight, and deep; hams heavy and well let down; bone fine; hair ordinarily rather thin; fattening qualities very superior. The breed is becoming very highly esteemed and very popular in the United States as a cross with the coarser breeds and more slowly maturing qualities. They are good breeders and fair nursers, always fat and healthy and hardy."

This description is remarkably close to the Guinea Hog. It could be that, when the Improved Essex was imported in the late 1800s, it was popular with farmers in the Southeast, crossed, not registered, and renamed as Guinea. Or the Essex could have been crossed with an established local breed known as Guinea. Crossing would explain some of the variety and types from "little-boned" to "big-boned."

In *The Livestock Conservancy News 2013* article titled "Pig Breed Relationships," D. Philip Sponenberg summarized a genetic analysis of pig breeds led by researchers in Barcelona, Spain. The

Ossabaw Island and Guinea Hog breeds, both North American lard hogs, were compared to Iberian pigs, Chinese breeds, and European breeds, specifically Latin American breeds including Duroc and the Black Canary Island breed. There were no English breeds such as GOS tested in that study. Sponenberg commented on the findings, "The Guinea Hog stands apart as the most distantly related American breed. Exactly how it fits into overall "hogdom" is still uncertain, but it does make the Guinea Hog stand out as a conservation priority."

Sponenberg commented elsewhere regarding the study. "The DNA testing of the Guinea Hogs sampled showed homogeneity, demonstrating that it is a pure breed. However, it was a clear outlier and not closely related to the other breeds tested...In the case of the Guinea Hog, it will be especially useful to track down any source herds that have been overlooked in the past—these will greatly help to provide the genetic strength that the breed will need to move into the future."

Gra' Moore, the breeder who provided the hogs for the Slow Food Charleston event, comes from a family with a long history of raising hogs. Gra's father was born in 1940 and remembers the hogs his father (Gra's grandfather), born in 1893, raised on the river in North Carolina. They had been Improved Essex. When Gra's father first saw the Guinea Hogs his son was raising, he asked him, "What are you doing with Essex hogs?" This anecdotal evidence indicates a similar phenotype between the Guinea Hog and the Improved Essex and aligns with the DNA evidence, or genotype, found in DNA analysis. Because the Essex are extinct, there is no direct DNA evidence, only the indirect connection in relation to GOS mentioned earlier in this section. Currently, staff members at The Livestock Conservancy strongly suspect a significant connection between the Guinea Hog and the extinct Improved Essex.

Chapter Two
Issues of Black and White (And Red or Blue all Over)

"Breeds serve as genetic resources because they have predictable combinations of genes throughout most or all individuals of the breed....Owners choose specific breeds because they are interested in a certain appearance, performance, and behavior. Predictability allows owners to be satisfied with their breed choice."
Sponenberg, Martin, Beranger 2017
Managing Breeds

"What is black and white and read all over?
A newspaper!"
Common childhood joke in the 1960s

Color Variations in Guinea Hogs

As indicated in the current and past breed descriptions, Guinea Hogs were predominantly solid black, with a rare variant of minor white on a foot or the snout. Due to inbreeding and human selection due to preference, white marks are becoming more frequent and predominant in certain herds with Setty, Celesky, and/or Baylis in the background. The GHA breed standard called for solid black, with more than a few white markings unacceptable. The AGHA breed standard (description) is: "Most American Guinea Hogs are solid black. A common variation due to a widely spread recessive gene, is solid black with minimal white points at the feet and tip of nose. Excess white (beyond the feet and the end of the snout) is discouraged. An extremely rare recessive red gene exists in the breed, and may rarely exhibit."

The "rare recessive red gene" alluded to in the breed description is rarely exhibited, although could be hidden under the dominant black gene, a friend explained to me. There are reports of piglets born red that later darken to black, or black piglets that develop red hairs as "highlights." I once owned a gilt that was born black, turned red almost overnight, and then turned black almost as quickly. None of the foundation hogs entered into the AGHA registry were red, and none had red highlights. Sometimes breeders announce that they have a goal to "bring back" the red bloodlines. I am not sure that they ever fully expressed in the American Guinea Hog breed. At least I have not found any evidence that it did. The ginger-born or red highlighted hogs described above eventually become solid or mostly black hogs.

Blue is a rare variation that is well documented although not mentioned in the breed description. It is often linked to hog strains that carry minor white, such as Celesky/Setty. These pigs are sometimes, but not always, born with a white spot on their nose. The skin beneath the spot is pink. At some point in their growth, they develop white hairs. These can radiate from the pink nose, if there is one, or white hair can sprout on other parts of the body. The white hairs spread and have a silvering effect when interspersed with black hairs. This can resemble a blue tint. Not all hogs that later turn blue are born with any white. Many, if not all, have blue eyes. Sometimes the silvering effect only partially covers the body, and sometimes it is so dense that the hog appears almost white.

Gra' Moore had a blue sow purchased from the Beardsley Zoo in Connecticut. In later adulthood the white hairs reverted to black so that the blue color is no longer evident. Founders in the AGHA with minor white (one white foot and a spot on the nose) include Baylis

Samson and Setty Houdini. The Celesky hogs were bred from Randy Setty's and possibly Donna Watkins' stock, so are likely related to Setty Houdini, Setty Rose, and other Setty hogs. Celesky and Setty hogs are linked to the uncommon blue variance, found in blue hogs at the Beardsley Zoo or their descendants.

Billy Frank Brown in Mississippi had a blue boar he obtained from Mr. Keene in Pitts, Georgia. An offspring of that boar, also blue, was given to Arie McFarlen of Maveric Heritage Ranch in 2005. She listed him in her herd as "Brown's Blue Boy." Descendants of Brown's Blue Boy were entered into the AGHA registry beginning in 2017 as part of their Genetic Recovery program. Mr. Brown reports that Blue Boy, when bred to a black sow, produced one out of four piglets that were born blue. They did not turn blue after weaning, as is more typical. A black daughter of Brown's Blue Boy, Maveric Esther, also produced a gilt born blue, according to a former owner of that pig. In a phone conversation in 2015, Arie explained to me that there are three different ways that the color blue expresses in the Guinea Hog. One is a blue coat, one is blue eyes, and the last is blue skin.

Color Controversies

The Guinea Hog breed, like all breeds, has a unique phenotype that expresses the genotype. As a landrace breed, the Guinea Hog has a broader range of acceptable phenotypes than a standardized breed. Its breed standard, you may remember, is written as a description rather than an ideal prescription. Those who choose landrace breeds find this variety appealing. Because variation is allowed, breeds sometimes have situations where rare, but acceptable, variants become more popular. When that popularity increases, as in fads, formerly rare variants can become predominant.

This seems to happen often with color traits, according to D. Phillip Sponenberg, who has written extensively on color genetics. When it does, members of associations may disagree on these changes, and this can cause contentious relationships within the group. Breeding for color fads is addressed in *Managing Breeds*. "The danger is always that fads will drive selection to favor what were originally rare variants to the extent that the breed changes character." This is balanced by, "Standards for landraces must allow more variation than those for standardized breeds."

It is important to understand the history of a landrace heritage breed and what our elders' practices were if we desire to retain particular lines from a historical perspective. It is also helpful to understand how breed descriptions were developed. Discussions around these issues are often heated, to put it mildly. History seems to repeat itself. My hope in this section is to provide historical background for current and future breeders.

Mostly solid black is the norm, but since the population's genes existed before the breed description was written, other variants may express from time to time, as discussed in the previous section. Old-timers in populations from Georgia and Mississippi assert that the blue color variant frequently occurred in the past. Some current breeders find this variant very exciting when it expresses in their litters. However, when it appeared at the Sedgwick County Zoo in 2008, many breeders were suspicious that crossbreeding had taken place.

Because blue hogs are born black and appear silver later, their color is listed on most of their AGHA registration forms as black—the color of the hog at the time of registration. The owners do not typically update these records, so it is hard to determine if a given hog in listed in a pedigree carries that genetic trait. You would have to do further research or purchase from a breeder knowledgeable about your hog's ancestors.

ISSUES OF BLACK AND WHITE

In 2015, D. Phillip Sponenberg wrote a summary of color traits in the Guinea Hog for members of the AGHA. The following quote is an excerpt from that summary.

"Guineas are generally black. They can and do produce some hogs with minor white marks on feet and head, and likewise the occasional belt. This is expected, due to the breed's heritage. Also, on occasion, some dark red or blue hogs have been produced from pure stock. These odd ones pop out because recessive genes still occur in the purebred hogs, and can hide under the dominant black for many generations.

"While some color variation can occur in Guineas, there are a few things that cannot just pop up out of nowhere in purebred Guinea Hogs. These include sandy red (ginger), red (ginger) with black spots, and sandy with black spots. Likewise, white with black spots is highly suspicious that crossbreeding has occurred. In cases where these colors occur in reputedly purebred Guinea Hogs, it is doubtful that the stock is indeed purebred. This is to the extent that in such cases the parents, and ideally the grandparents, should all be DNA verified to assure that the odd hogs are indeed purebred." (Author's note: Another color that occurs in crossbred Guinea Hog newborns is a striped coat. This usually indicates that genetics include Ossabaw Island, Mangalitsa, or Eurasian wild hog.

Breeders must choose between breeding deliberately to accelerate and maximize rare traits, accepting the occasional white foot on an otherwise superior hog, and minimizing and culling hard to eliminate traits that do not impact the health of the animal. In an online book discussion group, Sponenberg clarified that "saving an off-color but poor animal is a poor choice!" Making breeding selections with a landrace breed is tricky, and must be approached thoughtfully. Breeders are advised to think about where they stand on this continuum, think about their preferences in advance, and follow through when it is time to make their selections.

In a telephone interview, Dr. Sponenberg and I discussed breeding on a spectrum from no white to increasing white. He stated, "In general for rare breeds, there is an emphasis on some colors, but you are likely to choose other things when you eliminate a color. That being said, I'm not sure that breeding for whiter and whiter hogs is the way to go. If a breed description defines too much white, then a pig with white should be bred to a solid colored hog rather than breeding it to another hog with more white. But I am opposed in general to denying registration based on mismarking of color. You don't want to use a lousy, otherwise off-type black hog. But you do want a really good, typical hog with small white markings, if it has good traits related to production or soundness. Basically, all breeds need selection and culling."

The take-home message here is that certain color patterns are not likely to be found in a purebred Guinea Hog population. Without history to lean on, it is possible for breeders seeking novelty or cuteness in their herds to imagine a time when Guinea Hogs with four white feet were common and accepted. There is no oral or written history to support this. The elder breeders considered white markings to be signs of either inbreeding or crossbreeding. We have been seeing more white in recent years. This is likely a rare variant coming to the surface after fifteen years of heavy inbreeding. It could also be due to breeders deliberately selecting hogs that carry the white markings. In either case, it is not representative of historical type. I asked elder breeders about the colors of their hogs and their insights helped me understand the traditional perspective.

Historical Data and Traditions of Long-time Breeders

I talked with Kevin Fall, a founding AGHA board member, in 2014 about how people discussed the hogs back in 2004, when he first began to raise them. I wondered if they were considered basically an all-black breed. He shared his thoughts.

"Yes, that was my understanding of it. One thing that we always considered part of the reason that white shows up in the Setty pigs is because Setty Houdini and Setty Rose were DNA tested as being littermates. That may be the reason that showed up. If I have a boar that has it [white markings]' I will discriminate and butcher him. If I have a sow that has it, I will consider keeping her. A sow can only have two litters a year, but a boy can breed all the sows. So if it is an inferior trait, I don't want to transfer it on through the boar line. With the sow line, it depends on the genetics and how important it is to me to keep her. I look at the historical descriptions that we have.

"There were basically three colors - black, blue, and red. The Improved Essex breed description was for a small black pig. That is what I like and what we went by in the organization. And that is the direction that I have always watched, and the way I have tried to direct my breeding. We kept minimal white in the description of the hog because it was a landrace breed. It showed up every now and then. We only had twenty-six viable animals in the United States, so we started with what we had."

Kevin and I spoke again in 2017. I asked him if he ever got hogs with white feet or white noses. He explained again that the white seemed to come out of the Setty line due to heavy inbreeding. He made the following suggestion for breeders.

"If the breed is getting more and more white, or hogs with four white feet, I would be testing to see how much they are getting from the original lines of the pigs. Is it coming out because of linebreeding, because of selecting for color, or because somebody has thrown in a Pot-bellied Pig somewhere?"

I asked Don Oberdorfer about his advice for breeders in regard to color in the Guinea Hogs. He replied, "I think it is important not to be super distracted by sort of incidental cool things. I've had some pigs that turned blue after they were castrated. And by blue, I mean there were white hairs mixed in with the black that give it a bluish sheen.

When someone has a blue boar, everyone wants to breed them to be blue. And so wait a minute, with such a small population, we can't start tweaking them for surface characteristics and for one sort of attractive surface characteristic."

J. Frank Baylis in Tom's Brook, Virginia bred the only other foundation boar besides Setty Houdini that expressed the white trait. Jim and Shirley Sullivan of Sullbar Farm owned Baylis VA Samson in his later years. Shirley told me the boar had a single white foot and a spot on its nose—both of which were barely noticeable. Mr. Baylis had strong opinions about selection, and attributed the appearance of white to inbreeding, as have so many other breeders. J. Frank told me this about his closed herd.

"I'll tell you also what we got sometimes from being very inbred over the years, was a white cornet band on the back foot. I stopped buying outside pigs. I just inbred. White feet went to slaughter right away, because you don't want that. Colors, whenever you breed animals, you can change and make new colors with purebred animals. Over the years, colors crop up. That was the biggest problem that I had with breeding Guinea Hogs. Some people had health problems, but I had the cornet band. One out of twenty-five pigs would have it, only one foot in the back. A lot of times it would go away, and sometimes it didn't."

Kirk Fackrell of Cascade Meadows Farms drafted the breed standard for Guinea Hogs. He expressed his opinions about the role breeders should take in selecting offspring for breeding. He is concerned about increased white expressing in the Guinea Hogs. "One of the reasons that color is an important part of the breed description is that it becomes kind of a trademark. If people start to mess around with it too much, they can do anything. I can make these hogs giant if I had ten years to work on it or make them teeny tiny miniature ones. So the purpose of a breed description is to give people a breed standard

ISSUES OF BLACK AND WHITE

as a guide. If we don't all try to honor this guide, the breed will stop being a breed. It will be everything. So a lot of people think to have a breed, you just breed. You take purebred whatever and breed it to another purebred and all the offspring will be that. It's not true.

"You have to keep selecting and selecting. Otherwise, over time, it will be a mish-mash of everything. We had a big debate in the early organization about the white issue. We didn't want too much white. On our farm we wanted the white-footed ones occasionally. They helped us identify the pigs. But one of the reasons we talked about not wanting too much white on the pigs is that it put them into pet status. Then someone says, 'That's a special pig. I love that one because of the white, and I like the white.' And then we will have pet pigs. That was a huge part of the discussion from 2005 to 2007. We really wanted to avoid these pigs becoming popular as pets. Pet status will destroy them. And even now today, when somebody wants them as a pet, I say no."

He continued, "Instead of selecting for color and pet status, we should be selecting for those traits that have made it a unique breed for generations. Does it have parasite resistance, breed easily, and have good mothering abilities? Color is the enemy of form, because people will focus on the color and ignore more practical traits. It is totally bad for the breed when people get on color fads going after more white, blue, and so forth. About one-fourth of our herd has some white on it. Sometimes they will have too much white, and we will not sell it as registered. If we breed out of a male, they have to have minimal white on their feet. And then anybody who is drawn towards white, we go okay, that is fine if you want a few in your herd, but don't be selecting to have them all white footed. Because then you start to have breed confusion with Berkshires.

"And the idea is that if you said Guinea Hogs are solid black or solid black with a few white socks or white feet, then is that a

Guinea Hog? If their white goes up too much further who knows, it might be confused with Berkshire or if it goes up and makes a color it would be confused with Hampshire [hogs]. So we said that was a strong point of discussion. There were some people who said, 'It says solid black and we should stick with that.' I said, 'The problem with that is that there is a large percentage of the Guinea Hogs that have the white foot gene. If you said that was unacceptable, I would say that we should DNA test it and remove all pigs with that gene from the breed. And that would be bad for the breed because probably 40% of them have the gene.' So any way, we said, 'Well, let's write that into the breed description because that is what we have.'"

Kirk continued, "One of the powerful things about heritage animals is that they were able to survive and thrive in a time when we didn't have a lot of vets and chemical dewormers and vaccinations. Those things to me are probably the most critical selection criteria. If a pig is a little bigger or smaller, that is not as important as a pig that can fight off parasites with no chemical help. As a matter of fact, our rule is they can't require any deworming at all. Now that is just our farm. I wouldn't expect everybody to have that rule on their own farm. But on our farm, we go, 'If you need to be dewormed, we're going to kill you and eat you.' Another trait is the ability to maintain good body condition on just reasonable grazing and without supplements." (Author's note: not everybody's pasture provides reasonable grazing. Cascade Meadows has forty lush acres with healthy diverse grasses. If you have a weed-filled, unimproved pasture, please provide supplemental feed as needed for your herd.)

He added, "So that is the important thing that this breed does for us. It can save people's lives when times get hard. We don't need this breed to be a good market pig that you can sell to your friends to have a really great meal in an expensive restaurant. There are plenty of other pig breeds that can do that. That is not a core requirement of this breed. A core requirement of this breed is that when times are hard and people are struggling to survive and find food, this breed will be there to help people when times are hard. Now at the same time, if times are easy, sure, we can provide some to restaurants and what not, that is great. But the core requirement is a pig that is going to take a family near starvation and you can take it out feeding on grass and you can feed your family. Without chemicals and without a vet and without anything. Just some grazing and a few scraps to survive and thrive. That is the one role that Guinea Hogs can provide better than any other pig."

Gary Sumrall, a third-generation breeder in Mississippi, also felt strongly about keeping white from creeping into bloodlines of the Guinea Hog. Gary's family had been breeding for three generations, predating the registry by 90 years. When I asked him what he preferred if an otherwise good hog from his bloodlines was born with, for example, a white foot, he replied, "I think I would slaughter it. I think I'd kill it."

Shirley Sullivan was an early registrar for the AGHA in 2007. She told me that certain members of the organization considered the appearance of white "very suspicious." So when Shirley had a piglet born looking "like a Boston Terrier with a white belt," she kept that news to herself. She did not want to be the target of suspicion or attack, even though she knew that her stock was pure and not crossed. Shirley's husband, Jim, the second president of the AGHA after Don Oberdorfer left the board in 2007, spoke to me about his memories of the hogs and controversies about color, as well. "We would have different piglets that would have a red tinge at birth and then turn black. That was going way back early on, everybody said, 'Oh, my gosh, there is white in that one, that is not a Guinea Hog.' Well, it could be, because there could be a recessive gene that would come out. And then the Beardsley Zoo had one of the blue ones, at one point. Nice looking hog. And everybody was upset about it."

It is my hope that breeders of Guinea Hogs will consider various aspects of function, form, type, strain, and tradition in their herds when making selection choices. Color is just one factor. I encourage breeders to develop a breeding philosophy of deciding what you want in your herd, then choosing strategies and setting goals that support your philosophy. Put it all in writing and articulate it to buyers. Remember the main priority: selection, selection, and selection.

Part II
Raising Guinea Hogs 1940-1995

Chapter Three
Era of the Southern Hog: Interviews with the Elders

"Well, you know, it don't take too much to feed them. We built a big pasture for them in a wooded area where they could get acorns and run in the woods. And we just kept raising them. We bought a couple, and we just started keeping some of the females 'til we had probably sixty or seventy of them."
James Priest

Fewer Fences

Billy Frank Brown, born in 1942 in Mississippi, has raised landrace livestock breeds his whole life. These breeds, native to and adapted to the land, include Pineywoods cattle, Pine Tacky horses, Gulf Coast Native sheep, and local hogs. The animals were notched or branded to identify ownership. After that, with the exception of horses, they were allowed to run in the woods. "We earmarked the pigs," he told me. "Back then, the settlers, they respected one another and if they came across a sow that had a bunch of pigs, you know, if any of them weren't theirs, they'd mark the pig with the other fellow's marking."

Don Bundrick, born in 1954, grew up in Rabun County, Georgia. On a visit to my farm to see my Guinea Hogs, he recalled how farmers raised their hogs in the 1960s. "The farmers turned the hogs pretty much loose in the woods, especially those that had bottom land with acorns. Even when I was a grown man, my uncle would turn them into a cypress bottom. I asked what they were eating, and he said, 'Pretty much whatever they can find.' They'll eat frogs, anything they can find. And they are very creative in their browsing habits. I remember that after turning them loose all summer, people would round them up in fall and winter. In Rabun County they were raised in the woods and driven across the river into South Carolina, where they were traded for vegetables. The next county had huge vegetable production until it transitioned to apples. I remember talking to some of the old timers. They told me that as children and young men, they drove hogs across the river and down at the very end of the Chattooga River into South Carolina, where they would trade."

Noah Maloy's family has lived in Jefferson County, Florida since the mid-to-early 1800s. His grandma's grandfather was born there, and the property has been in his family for generations. "Growing up," he told me "We didn't have electricity, and we didn't have indoor plumbing or anything like that. So that is the way I grew up until the late 1960s. What we had, as part of making ends meet, was our hogs, and our hog claim. A hog claim means that you have a registered mark that you put in your hog's ear. The mark is registered at the courthouse. Then what happens is, you pay taxes on the amount of hogs that you claim you have. If you claim that you have twenty-five hogs, then you pay taxes on that. And people always had more than they claimed they had. The way it worked for us, is that our hogs mainly went between the Saint Marks River and the Wacissa River. And the cows, too. It wasn't just hogs, it was cows. In the early days, and before I was born, in that part of Jefferson County, you didn't have any roads or any highways. What you had was old railroad beds, and you had kind of like pig-trail roads that went through the place. They didn't just have Guinea Hogs. A lot of what we had were called 'Pineyrooters.' They were a little bit different body type than the Guinea Hogs.

"They would run wild, and a lot of them would stay around home. Out at the old corn crib where the barn was, you could go out there, and there was a hole cut in the door. There was a chain that went through it, and you hung it on a nail. That was the latch for the door.

You would go out there, and you could rattle that chain in the evenings, and the hogs would come out of the woods and come up there. And you would throw them some corn. By doing that, they kept coming back. And that was the main reason you did that.

"The Guinea Hogs are really good grazers. I remember the old people calling them either small-boned or big-boned Guinea. I don't know how you got one or the other. Maybe they were just distinct body types. I don't know. I was pretty young back then when my grandpa was living. I grew up in a time when he was like eighty-six years old. But from the time I was little, I remember him going out and getting his horse. He would fill his saddlebags full of corn, and he would go out in the woods. At certain spots where the hogs were, he would throw out corn. And when I was growing up, I would go out on horses, too. I would catch hogs and cut (castrate) the boar hogs and mark them. The old people would cut the tail off a barrow so you would know he wasn't a boar hog. A lot of the old people talked about Guinea Hogs. They were easily kept up. You didn't do much with a hog."

Butchering Time

Billy Frank Brown told me about butchering time at Cowpen Creek. It was a community affair.

"We cooked with the lard, and I make crackling every year now, at Easter Sunday. A big pot of crackling. And we used the lard all year to cook with. It won't kill you. They say, you know, it ain't good for you, but my mama lived to be ninety-four, and she ate salt meat every day. Every morning she was going to have some salt meat — bacon, you know. The old-timers, they lived a long time.

"Lard makes good biscuits. My wife makes biscuits with it, and I cook out about a gallon every spring of the year. We use it all year long. We used to make sausage and cook it and store it in the lard to hold during the summer. They didn't have refrigerators back in those days, way back in years."

I was fascinated with the story about covering sausages with lard to keep at room temperature with no air conditioning in the deep South. My mother grew up in Illinois in the 1930s. A few months before she passed away, she told me about how the neighbors came over to help her father process a hog. My mother's job was to pour the lard over the sausages. Then they were stored in a hole in the ground. "Can you imagine?" she asked me. Well, now I can. I heard two or three versions of this process in my interviews.

Billy Frank continued, "You cook meat, your meatballs and your sausage balls and they'd just keep in that lard, you see, in the pantry. You had to put it in quart jars and just pour it in there and save it, you know? It's there until you get ready for it, you know?

"Usually from the smokehouse, you'd smoke sausage. They didn't last long. They'd keep them and the bacon on the rack in the smokehouse. We ate them up fast. Usually, we didn't have much pork during the summer. We'd have a hog killing in the coldest day of the year, like Thanksgiving or Christmas. Kill about eight hogs and smoke the meat, you know. And it'd last until up in the spring of the year."

He explained, "Everybody and his brother came and helped. They'd go to other people and other places and do the same thing at their place. You'd have a big hog killing. That was in the 1950s. Then they started coming out with these deep freezers, you know. The people started putting their meat in the deep freeze. I can remember we didn't have a deep freezer until sometime close to 1960. They'd already come out with the refrigerator, but I can remember the first deep freeze we got."

Don Bundrick also shared memories of butchering time in the 1950s and 1960s. "Groups of families butchered together to make it easier. Generally, the hog was shot in the head, dragged over to a big syrup cauldron, and scalded and scraped to get the hair off. Then I remember they hung them in the trees. The women took over then, with their sharp knives. They would render everything on that hog. Everything. My grandmother was famous for her sausage. There was a rumor that her sausage was something that people would come over and get her recipe for their business. That is a sore subject for some of them."

James Priest, born in 1934, had a long career at Lockheed. When he retired he continued to keep a four acre organic garden. He was still maintaining it at age seventy-nine, but slowed down a bit after a traffic accident. James had not seen a Guinea Hog in over forty years until he met one at a friend's house in 2015. By 1954, they were just that uncommon. James became head of the household at an early age. He was only fourteen years old when his father passed away. He supported the family by selling Guinea Hog feeders, raising rabbits, growing out bull calves, and working at a dairy farm when he was not in school. He remembered hog butchering.

"When it came butchering time," James explained, "we would go from house to house and start butchering each other's hogs. We would butcher about twice a year. We had all of the provisions when it came to butchering them. I was just a kid, when my daddy was living. We butchered them all by ourselves at our house. But when he passed away, I had help from other people in the community at butchering time.

"I had a smokehouse. My daddy had one, too, and we used to keep it full. It was huge, I guess about a sixteen by sixteen foot smokehouse. Our hams and shoulders were salted down. They stayed in the salt for so long and back then you would cover them up with salt. Then you would take it out of the salt, wash it off, and hang them up. A time or two, in my daddy's later years, we did sugar cure a couple of times. I liked the sugar cured hams."

James continued, "My mother took and made pressed meat, souse meat, out of the head. That is good stuff. We had a colored family that lived pretty close to us. Him and his wife would always come and help us butcher, just to get the feet, ears, and anything else. They were such a good couple and good workers, you know. That is just a great family. Great neighbors.

"We kept the lard. We would cook it out. My mother would trim the fat meat off, and we would cook that stuff out, and can it in half gallon cans. And there would be crackling. We'd make crackling bread. It is good. You put it in the corn bread and cook it. It is a little greasy, but good. We would grind the sausage, and my mama would can it in a jug and set it up on the shelf upside down to let the lard seal the sausage off and keep it fresh. So when we wanted sausage, we would go and get a can."

Freddie Brinson, born in 1950, lives about fifty miles south of Augusta, Georgia. He spent thirty-one years teaching history and had served on his local school board for seven years when we spoke in 2015. He talked with me about blue Guinea Hogs the family owned, and about butchering time.

"When I was growing up, we slaughtered the hogs at home and almost everything that could be used, was used. It was kept. But I can still smell the hog. You had to put hot water on it to get the hair to come off. After you have killed them, you know, either you've got a big vat to put them in, or you pour hot water on them and the hair comes loose. I can still smell that smell from the hot water being poured on to get the hair loose and scrape the hair off. We did it when I was very young, but my parents had been doing it for years, all the way back to that little blue Guinea Hog sow in the 1930s. That was the first one they had after they got married and started raising a family. We always slaughtered in the fall.

Freddie said, "It was common for families to have a smokehouse. Everything was pretty much processed and then put in there. They would hang for a long time. The meat tasted so different. It just had more of a taste to it. Of course it was always, even after it cured, it had a fresher taste than meat today. Just a completely different taste.

"As time progressed, by the time I got to my late teens, I had my own little blue Guinea Hog sow. We had stopped butchering at home at that time. We took them to the butcher. We have a local slaughterhouse here. But we stopped doing that in my late teens, right before I started college. I remember that we would load them up and take them to be slaughtered."

These stories illustrate how valuable the Guinea Hog was to families throughout history, how it kept the families fed during the cold winters, and how neighbors came together to help each other in the process. The disappearance of these hogs began without most families realizing the loss until it was almost too late to save them.

Part III
Why did the Guinea Hogs approach Extinction?

Chapter Four
Vanishing Hogs

"They disappeared, because people's demand
was for leaner meat and farmers were
moving toward industrial farming."
Don Bundrick

"Then they got to breeding the longer,
thinner hogs and the meat just wasn't as good."
Cohen Archer

Changing Times

According to The Livestock Conservancy, about half of the swine breeds existing at the beginning of the twentieth century had disappeared by its conclusion. There were several major changes in agriculture that contributed to this breed loss. Consolidation of agribusiness into fewer, larger farms was one factor. Consumer demand that favored lean meat was another. In addition to those factors, the pet pig craze impacted the Guinea Hog breed specifically.

Don Bundrick recalled, "I can't tell you how much it changed. Just the average farm changed. When I was a kid in Rabun County, Georgia, you never saw a field that was more than five acres. But once the advent of irrigation began, you had these giant fields. It moved things toward an industrial type of agriculture. To me that has taken so much of the good things about farming out of it. You would have thought, after the Dust Bowl era, that people in the South would have looked up and realized the problems.

"I remember that all of a sudden, after the 1960s, the hogs changed. They became larger and leaner. And for some reason, peoples' tastes changed. Somebody convinced them that leaner pork was better for you. I am sure it had something to do with the medical community, but I don't know. Anyway, the hogs got harder to handle, and their demand on the land was harder. My dad always told me, 'You be careful with these larger hogs, because they don't have the good nature of the original hogs that we used to keep. Those, we could let them run, and you didn't worry too much about your kids or whatever. These hogs, if they knock you down, they will eat you up.'"

He continued, "The Guinea Hogs were, from what I understand, pure bloodlines. It is amazing that in such a short time they disappeared. Because, I guess, peoples' demand was for leaner meat and farmers were moving toward industrial farming. When we were kids, those fence rows would grow up with trees. We would quail hunt there, and then we were working with smaller tractors. Generally there were a lot of paddock-sized pastures. Cows would rummage in there if a crop failed. The wind blew the corn down in the fall, and they would turn the cows and hogs in there.

"At the turn of the century, all the trees had to be cut over. The National Forest was instituted because the waters were running red from runoff. There was no overstory to protect it. The same people who bought the property and sold the trees to the logging company were the very same people that turned around and sold it to the national government for the National Forest. So the government became stewards of the property until it grew back up."

After speaking with Don, it was even clearer to me that industrial farming was a huge factor in the changes that nearly caused Guinea Hogs to disappear completely. Don continued to describe this era and the mindset that led to a decline in the Guinea Hog.

"Then, due to pressure from the universities, manufacturers, and lumber industry, foresters converted to monoculture practices that led to things like pine beetle infestations and problems with hemlocks. We are still dealing with these issues today. Until we get to the point that we recognize that bigger is not better, things will not change. People need to be allowed to make a living wage from farming.

"The average age of a farmer is sixty years. When do you hear about young people interested in it? I can remember my grandparents and other relatives tell me that during the Great Depression, they had a house to live in, and a garden. They were so much better off than most, even though their cash money disappeared overnight. A teller window was literally shut in their face and they had nothing, but they survived. They never trusted banks after that, and never bought anything on credit. If they needed something, they would do without until they could pay for it with cash.

There should be a living to be made in farming, but it isn't for everybody. People have no idea where their food comes from. They eat meat, but oppose getting their hands dirty with it. With the hogs particularly, I think it is a great idea to have the Guinea Hogs reintroduced to farms and into our diets. Without fat, pork has no taste at all."

About a year or so after this interview with Don, I helped him support his son's wishes to be one of the young people to revive some homestead skills. Don gifted Nathan with a trio of breeding Guinea Hogs from some rare genetic lines to start his own herd. They went to Taterville Farm in Tallulah Falls, Georgia, an area where the Guinea Hogs once roamed in the woods. Their meat fed the local smallholders and nourished them with meat and lard during the cold winters.

Numbers of smaller farms such as Taterville Farm are on the increase, but are still the exception in the world of industrial farming. A major factor in the increase of industrial farming was government policies that helped bring long-held traditions in family farms to a halt.

Earl Butz was the Secretary of Agriculture under President Nixon and President Ford, serving from 1971 to 1976. During this time, he implemented dramatic changes in United States agriculture. His policies have been lauded by some, and heavily criticized by others. He abolished programs implemented that reduced intensive farming after the Dust Bowl era (1930-1936) that had subsidized farmers for resting their land. Those practices rewarded farmers that produced fewer grains. They also kept prices high enough for farmers to be compensated adequately. Small farmers had supported their families and their lands by diversifying and receiving compensation that covered costs.

Butz ended those policies. He urged farmers to plant commodity crops from "fencerow to fencerow" and challenged them to "Get big or get out." Farmers began switching from diversified farming to a monoculture model. When farmers planted more grains than they could sell domestically, the excess was sent to the Soviet Union. Policy shifts implemented during this era coincided with the rise of corporate agribusiness. As food prices dropped, financial stability for small family farms declined. It became difficult to compete with the larger family farms and corporate-owned factory farms.

This corporate-owned trend has drastically affected the pork industry as well.

Consumer Demand and Industrial Strains

Throughout history for generations, fat was considered highly desirable. It held positive connotations, not negative ones. Fat was useful in many ways and held in high esteem by those who consumed it. Wealthy, powerful community members were those who had enough food to store fat on their bodies, therefore, plumpness was equated with beauty, prestige, and wealth. All of this changed in the mid-1950s, as it was determined that heart disease was a leading cause of death and the associated lipids theory of heart disease was widely accepted; though never proven. In fact, as lard consumption was replaced with trans fats and polyunsaturated oils, incidences of heart disease steadily increased—not decreased.

Today, many people are adopting a ketogenic or paleo high fat diet to lose weight, reduce inflammation, improve brain function, and improve their lipids profiles. Clinical data can be found on both sides of this issue. I have a personal bias to eating real foods and traditional foods over highly processed foods full of chemicals and preservatives. That preference is my own, based on information gathered over thirty years of personal study. I am not a nutritionist, however, and I am not giving nutritional advice. The information I provide about food is based on my opinions and serves primarily to contextualize the role of fat in American culture and the role of the Guinea Hog as a supplier of fat past, present, and future.

Fats, as eaten by humans for thousands of years, were found in natural sources. Families and home or producers in the village easily extracted the fats using animal labor, human labor or simple presses. The fats included lard, butter, tallow, olive oil, coconut oil, duck and goose fat, kidney fat, and palm oil. These fats are largely saturated and monounsaturated. Saturated fats are also found in animal sources such as meat, eggs, cheese, milk, yogurt, and kefir; all nutrient-dense foods. The same fats used for cooking and flavoring our foods were used in the making of soap, body lotions, and salves. Our forebears thrived on these natural fats and our bodies recognized what to do with them. These are the fats I choose to eat and serve to my family and friends.

With the government push initiated by Earl Butz in the 1970s to transition farms to commodity growing, farmers were now subsidized for growing corn, wheat, and soy. To keep this system going, there needed to be a market for the excess of crops. Newspaper and magazine articles, government pamphlets, and school curriculum included new recommendations. These included substituting polyunsaturated fats for saturated fats, decreasing consumption of saturated fat and foods containing cholesterol, and reducing fats while increasing carbohydrate consumption. Slick advertising campaigns convinced consumers to choose margarine, shortening, and highly processed corn and soy oils. These are fats made in factory settings and cannot be duplicated in the home kitchen. The solid fats are trans fats, now linked to heart disease.

Farmers were now growing more and more corn, wheat, and soy. Only so much of it could be exported. Food companies began to develop new products in order to take advantage of the low-priced subsidized crops. US Department of Agriculture (USDA) food guidelines still encourage "healthy grains" as a basis of our diets. Recommended servings have grown from two to four to six to eleven over the years. Chemical-laden, highly processed foods were made from grains and soy.

Much of the shelf-stable food was in the form of "fat-free" foods that were produced for consumers convinced that eating fat made them fat. To make these foods palatable, however, the fat was replaced with large amounts of sugar or fruit juice concentrate and gums such as guar gum and xanthan gum. Low fat food was in high demand, and health was negatively affected because the increase in sugar impacted blood sugar and insulin resistance.

Another way to sell the excess corn, wheat, and soy was to feed it to livestock. Cows and pigs were removed from green pastures to fatten in Confined Animal Feed Operations (CAFOs). Over the last sixty-five years, hogs have been bred to meet consumers' demand for less fat. The body types of modern breeds and particularly industrial strains raised in confinement are mostly muscle, not fat. Diana Prichard explains the problems with muscular meat in her article, "The Rise and Fall of the Great American Hog," in Modern Farmer, August 21, 2013. She laments that, "What started as a PR campaign for pork resulted in bland, unforgiving meat that upon cooking doesn't render enough fat to keep a layer of lipids on the pan."

"The other white meat" is now dry and tasteless when compared to the fatty heritage breeds and lard breeds such as the Mangalitsa, Ossabaw Island, Meishan, Choctaw, and Guinea Hog. The taste, moistness, and wonderful mouthfeel in the pork of these lard breeds comes from the fat, not from gum additives. The fat cover is what allows these hogs to endure cold weather and live in the woods and on pasture.

Chefs tell me that fat is a delivery method for flavor. The fat layers on heritage hogs deliver rich flavor and moisture to meat that is absent in modern production hogs. For the livestock, fat provides protection from wind, rain, and cold. Lard breeds such as the Guinea Hog can spend much of their time outdoors year-round. They have built-in protection from sun and cold.

Industrial strains of swine are owned and managed by multinational corporations, according to *Managing Breeds*. The largest such company is Smithfield Foods, Incorporated and is now a subsidiary of WH Group of China. The industrial strains are highly productive, but also highly dependent on human management, feed inputs, and electricity. While the Guinea Hogs are more unpredictable and low on production, they are rather self-sufficient and forage for much of their food. Conversely, the industrial strains are highly predictable and high on production, but very dependent on their caretakers. A single disease in crowded conditions can wipe out a large population of piglets in just a few days' time.

According to Barry Estabrook in his book, Pig Tales, 97 percent of pork consumed in the United States is now grown in CAFOs. In these settings, managed by corporations, millions of hogs each year are born, fed, and slaughtered in an environment where they never touch dirt or grass, never experience sunshine and mud, and live their entire lives on concrete. They live in close and constant proximity to other pigs and their movement is restricted.

Because of these conditions, workers dock the pigs' tails to prevent them from being bitten off and cut the eyeteeth of newborn piglets. The piglets have no access to dirt, so are given iron injections.

Antibiotics are added to feed to deal both with constant infections and to help them grow as fast as possible to finished weight. Indiscriminate use of antibiotics in agriculture may have contributed to the development of superbugs resistant to antibiotics in human and livestock populations, although this topic is debatable. At the time of this writing, policies are changing in regard to including antibiotics in livestock feed unless prescribed by a veterinarian.

As pork production evolved into factory farms and CAFOs, industrial pigs of today became leaner and ill-prepared for an environment outside of the climate-controlled setting. Guinea Hogs still have that protective layer of fat that allows them to survive and thrive in many climates. Industrial pig strains are delicate hothouse flowers compared to the hardy Guinea Hog. Today's farmers aim to raise the most pork in the shortest time on the smallest amount of space necessary. As a result, these CAFOs use more electricity and resources and produce dangerous levels of concentrated waste.

Genetic diversity from the rare heritage hog breeds could save the species if disease wiped out the industrial strains. If public demand shifts again to embrace fattier pork and humane pastured production where pigs can express their natural traits, these genetics must be preserved. Those genes have been lost in large part in modern breeds and industrial strains. Conservators of heritage livestock breeds, just like conservators of heirloom seeds, recognize the importance of genetic diversity in our food supply. A subset of livestock conservators has devoted considerable time and energy protecting the Guinea Hog in particular. These efforts are truly necessary as we learn what the small breeds and their conservators are up against.

Decline of the Family Farm

Using data from the U.S. Department of Agriculture's "Agricultural and Food Statistics Report" reporting on the 2012 Farm Census, Roberto

A. Ferdman of the *Washington Post* made some revealing claims. Ferdman quotes from the report, "Today's farms are fewer and bigger." The peak number of farms in this country was 84 years ago, in 1835. At that time, the United States had six million farms. At the time of the 2012 census, that number was reduced to only two million. In 1900, forty-two percent of families in the U.S. lived on small farms, many of them diversified. Since the 1990s, less than two percent of American families live on farms.

Farms are being sold and consolidated at unprecedented rates. Ferdman states, "the big farms have only gotten bigger over the years. As of 2011 — as is true with much of the country's wealth — the vast majority of America's farm land was controlled by a small number of farms. The top 10 percent of farms in terms of size account for more than 70 percent of cropland in the United States; the top 2.2 percent alone takes up more than a third."

The twenty years between 1950 and 1970, when the Guinea Hogs went from plentiful to scarce, correlates with a corresponding period marking a huge decline of family farms. There has been a recent resurgence of small farms and very large ones. There are fewer and fewer mid-sized farms, however. The median farm size in 2011 was 234 acres. Half of all farms have 45 acres or less. Thirty-four percent of cropland is on farms with over 2,000 acres. That leaves only sixteen percent of farms with 46 to 2,000 acres.

Fortunately for the Guinea Hog, it has a special role to play on small farms from two to 45 acres. It can thrive with little attention in such a setting. There is still a place for this niche animal on the homestead, and on a small diversified farm.

The Pet Pig Craze

Vietnamese Pot-bellied Pigs were imported from Vietnam in 1985, beginning a popular fad that continued into the '90s and still has a following today. These small pigs, similar in size to a little-boned Guinea Hog, were often crossbred with the Guinea Hog in an attempt to produce a friendly, miniature type. In several communities, city councils made exceptions to livestock regulations and began considering small hogs as pets rather than livestock.

With fewer people living on farmsteads and homesteads, all heritage hogs including the Guinea Hogs declined. Due to a push for low fat diets, lard hogs specifically decreased. With numbers of Guinea Hogs lowered and then crossed with Vietnamese Pot-bellied Pigs, there were even fewer of the breed remaining. The population was in dire straits, indeed.

It would take a small group of dedicated volunteers to turn the Guinea Hog population around and begin tracking, registering, conserving, defining, and promoting the breed. Only then would the tide begin to change back in favor of the humble black Guinea Hog.

Part IV
Organizing to Save the Guinea Hogs 1990-2006

Chapter Five
Getting Organized

*"Everything was run through Don Oberdorfer.
He did everything. If it wasn't for Don and
his work saving the Guinea Hog breed,
there would not be any Guinea Hogs left."*
Kevin Fall

*"People don't realize the effort and energy that everybody put in,
early on, to save the breed."*
Jim Barnett

Guinea Hog Breed Organizations Past

By the end of the twentieth century, Guinea Hogs were becoming scarce, and there were attempts to create an organization of breeders for the conservation of the breed. My research led me to an article that documented a Guinea Hog Registry active in 1990. It was titled, "African Guinea Hogs Make Great Pets," and found in *Farm Show Magazine*. It included an interview with Gary Spencer. The reporter wrote, "[Spencer] owns nine purebred Guinea Hogs and has started a registry for purebred members of the species." The reporter detailed the breed description for the registry as provided by Gary, who lived in Webster, New York. "African Guinea Hogs come in two types—small-boned and big-boned. At full maturity, the small-boned type is 20–22 inches tall and weighs up to 125 pounds. The big-boned type range is 30–36 inches tall and weighs up to 300 pounds. Both types have a shiny black coat and small, pointed ears....The big-boned variety can weigh up to 400 pounds if overfed."

An accompanying photograph shows what looks like a fairly typical modern Guinea Hog piglet. Mr. Spencer sold the males for $600 and females for $800. With inflation, those prices would be $1,125 for males and $1,500 for females. In 2019, registered, weaned shoats for breeding stock sell for $150 to $300 each, with mature proven stock selling for up to $600. In 1990, however, the Guinea Hogs were often considered rare and exotic pets, as the article's title implies. They were priced accordingly.

Gary purchased his herd from a man in the southern United States who had raised them for over eighty years. This article establishes the hogs' existence in the South as far back as 1900. I attempted to locate Mr. Spencer, but unfortunately, found his obituary instead. He did not live past middle age.

In 1991, the AMBC had been monitoring Guinea Hogs for about four years. They included a profile of the breed in a softcover book titled *American Minor Breeds Notebook.* The notebook, compiled by Laurie Heise and Carolyn Christman, was published by the AMBC and was "Made possible by generous support from the C.S. Fund and Jennie R. Donaldson Charitable Trust." The C.S. Fund was part of the Carl Stuart Mott Foundation.

Shan Thomas and Hans Peter Jorgensen of the C.S. Fund encouraged Gabriella Nanci, a teenaged member of the AMBC in Valley Center, California, to get involved with the Guinea Hogs. Gabriella had received a donation of Navajo-Churro sheep from the C.S. Fund to use in her 4-H club. Gabriella was keenly interested in rare breeds and was a very bright and motivated young lady. Gabriella and Lola Moffit, also from California, founded the Guinea Hog Association (GHA) to help track and promote the breed. They each lived in areas of California where the pet pig craze was popular. As high-priced Vietnamese Pot-bellied Pigs found their way into more homes and backyards, Guinea Hogs, often marketed as African Miniatures, started to receive attention as a cheaper and hardier source of miniature pig stock to cross with the Vietnamese Pot-bellies. Lola and Gabriella were concerned that the Guinea Hog would be absorbed into the Vietnamese Pot-bellied type pig, and the breed would be lost.

They worked in conjunction with staff at the AMBC. Staff member Carolyn Christman and executive director Don Bixby advised the GHA to set up an informal recordation of the Guinea Hogs, rather than a formal breed registry, and to develop a breed description rather than a breed standard.

Recordation for the GHA included diagrams that outlined hogs of three types with different noses, head shapes, and stature. Type A had a light bone structure and longer nose, Type C had a heavier bone structure and shorter nose, and Type B was a blend of the two strains with a medium dished nose. The GHA listed unacceptable traits in their breed description, making it slightly more prescriptive than the standard used later by the American Guinea Hog Association (AGHA).

Gabriella and Lola produced several newsletters that included information from other breeders, a breed standard found in the appendix, and newsy articles about the hogs. Only a handful of breeders paid the five-dollar fee to actually join the association.

Before 1994, Gabriella was not aware that Gary Spencer had started a registry, known as the Guinea Hog Registry (GHR). Gabriella told me that if she had been aware of it, she would not have started the recordation service. In 1994, Gary sent out glossy flyers to everyone on the GHA mailing list. He also posted ads that included a professional photograph of a Guinea Hog piglet with a white spot on its nose. Up to that point, Gabriella had never seen anything other than a solid black Guinea Hog. Gary told her he bred to accentuate white markings on the hogs.

Gabriella felt that a professionally marketed registry would stimulate more interest than her modest newsletter and recordation service. She therefore discontinued the GHA after two years. I was unable to locate any information on the GHR outside of that one article and my interviews and correspondence with Gabriella. There is no documentation of the GHR in The Livestock Conservancy archives, either.

Today, Gabriella is a bright and industrious woman. She is now a medical doctor and president of the nonprofit The Yonkofa Project. This project builds sustainable clinics in rural Ghana. Gabriella remains a member of The Livestock Conservancy, and continues to raise heritage breeds, although now in the state of Georgia.

In 2006, Gabriella gave Arie McFarlen of Maveric Heritage Ranch the GHA recordation files. This was just a few months after Arie obtained her Guinea Hog herd—in the fall of 2005. Arie was starting her own organization, called the American Guinea Hog Stewardship. Like the Guinea Hog Association and the Guinea Hog Registry, it never quite got off the ground. However, several breeders I interviewed mentioned Arie's registry. Arie maintained records of her personal herd and their offspring using reliable pedigree software.

Organizing the American Guinea Hog Association

Don Oberdorfer began his employment at the Dodge Nature Center (DNC) in 1998. The nature center is part of an educational institution in Saint Paul, Minnesota. He continues today as manager of the nature center and head of the Guinea Hog project, but the hogs are owned by DNC. From the beginning the nature center chose genetic conservation as part of their mission. Don invested much time and effort into seeking Guinea Hogs to start his conservation project. It was a challenging goal.

At first, Don was able to locate only two breeders. They were Mark Celesky in Cerasco, Nebraska and Bill Biggers in Bumpass, Virginia. Don also contacted Gabriella Nanci, who kept the GHA records, to help him locate breeders. It was through Gabriella's files that Don found Paul Krumm and his wife, Erin M. Taylor, known widely as "Micki." Paul had bred Guinea Hogs before he met Micki. Around the time of their marriage, they got a middle-aged boar from Gabriella Nanci. The Nanci boar was out of stock from Rosa Harper in California and Jean Cox in Kentucky. Those genetics are no longer in the gene pool. It would be amazing if anyone in those regions could track them down. By Don's recollections, the Krumm's stock had become old and stopped breeding, so Paul and Micki wanted to begin breeding Guinea Hogs again. They would be among the first breeders in the new association.

The first two Guinea Hogs that Don purchased were Celesky Tulip, a gilt from Mark Celesky, and a three-year-old boar from Bill Biggers named Arthur. Don picked up Arthur when returning home from a family reunion in Myrtle Beach. These two original Dodge Nature Center (DNC) hogs had their first litter on April 1, 2002. Don saved one boar out of that litter, DNC Junior. As of 2016, Junior was still alive and residing at Dodge Nature Center. He is phenotypically similar to his sire, Arthur.

Don traded two of Junior's littermates to Randy Setty in Goshen, Ohio in exchange for the adult hogs Setty Rose and Setty Houdini in September, 2002. Randy Setty lives in Goshen, Ohio. Goshen is a township east of Cincinnati and Don was attending his cousin Laurel Oberdorfer's wedding in Akron. So he did what all serious livestock conservators do when traveling anywhere. He arranged the acquisition to coincide with the family event.

He stayed over night in Norton, Ohio, with the brother of the bride, cousin Carl, and Carl's wife, Carmel. Carmel was also interested in raising rare livestock, so Don drove from Minnesota with a Fjord horse and Shetland sheep for her. The piglets destined for Randy also spent the night in Norton. This made for an eventful time with so many animals in tow and a wedding to celebrate.

Another DNC Junior littermate went to Stephen and Hollie Brothers in South Dakota. This was the renowned boar, DNC George. George had been a bit of a tourist attraction at Dodge Nature Center. With his laid back, easygoing personality, he was popular with visitors. Staff there would walk him on a leash and according to Stephen, the children liked to pet him. George was beloved by all three people who cared for him over the years. He is currently living out his life as a retiree, age seventeen at the time of this writing. The Brothers also got DNC Esmerelda and her littermate DNC Chunky, sired by Arthur out of Rose, and DNC Gabriella, sired by Houdini out of Celesky's Tulip. The Brothers were starting with a good sized herd for that time period, and would multiply quickly.

With four litters under his belt, Don was seeding and distributing the gene pool far and wide. He also promoted the breed in other ways. Don Oberdorfer had purchased a breeder's software program and kept a registry of the twenty or so pigs that he had been able to locate. He had been tracking them and their offspring for about two years. The registry documented his own herd, that of the Brothers, and hogs kept at the Sedgewick County Zoo by Callene Rapp, among others. This information would be invaluable going forward. However, one of the most powerful things Don did was to promote the breed through the written word.

In the fall of 2004, Don wrote an article for *Small Farmer's Journal*, 112th issue. It was titled "In Praise of Guinea Hogs." He covered some basic information about the breed, including its landrace status. Kevin Fall in Iowa was one of the people who read the article and reached out to Don to find out more about the hogs, probably in 2005. Soon after, Don told me, there was "a heritage pig insemination class that happened at the University of Missouri in conjunction with the sperm freezing project that the government was doing in Colorado."

Kevin Fall, Paul Krumm, Micki Taylor, Don Oberdorfer, and Donald Bixby, executive director of the ALBC, all met up at that course in Missouri held in August 2005. At dinner afterwards, with encouragement from Donald Bixby, they decided to form the official organization that would become the AGHA. Their intention was to have a democratic group that was run for the benefit of all members. They would keep a permanent registry that would belong to the group rather than an individual.

Following eight months of planning and promotion, Kevin Fall, Paul Krumm, and Don Oberdorfer signed the Articles of Incorporation for the American Guinea Hog Association with the state of Minnesota's Secretary of State on April 21, 2006. It had the following stated purpose:
- To promote interest and education about the American Guinea Hog to both attract new breeders for the conservation and well being of the breed and to promote its many uses.
- To protect the genetic diversity of American Guinea Hogs.
- To register and keep pedigree records of all animals qualifying as American Guinea Hogs according to the guidelines of the AGHA
- To provide technical support to a network of breeders to further their work in conserving the American Guinea Hog.

Formalizing this group as an organization and starting the registry was a crucial step in preserving the Guinea Hog breed and in saving it from its decline toward extinction. Kevin Fall summarized the situation as it started out in 2006.

"There were fifty-five Guinea hogs alive that we knew of in the United States. Of those, twenty-six of them were viable animals that could be bred. That means the rest of those, about twenty-nine, were sterile or too old to reproduce. There could have been more, but zoos and different places had castrated the animals. So we lost the genetic potential to begin with. But you know, if you started with twenty-six, you can tell people we have gone a long ways."

The process of preserving the Guinea hogs for future generations had begun. The three men who were founders served as the first Board of Directors of the AGHA and began to implement their purpose and vision. Just a handful of additional breeders registered hogs in the first year. Jim Barnett told me that when he assumed presidency of the board in 2007, the AGHA had a total of eleven breeder members. Since that humble beginning, the AGHA and the Guinea Hogs with the help of the AGHA have come a long way. There are somewhere around 300 farms raising registered Guinea Hogs as I write this. As of January 2019, more than 9,000 hogs have been registered through the AGHA. That organization has now celebrated thirteen years of operation.

Foundation Hogs of the AGHA in 2006

Foundation animals, in the case of a recognized and registered population, are the earliest animals recorded in the herd book that contribute genetic material to the breed. They are sometimes referred to as "founders." Barrows, sterile animals, and those without documented progeny are not considered to be founders because they did not contribute their genetics to the population. It is more beneficial to a rare breed to have a large number of foundation animals than a small one. A best-case scenario includes unrelated family groups contributing to the gene pool. In the case of rare heritage breeds, particularly landrace types, the situation is more often less than ideal.

By 1995, there were few Guinea Hogs remaining. Males kept for a long time had a chance to influence the future herd genetically more than the females did. A male could sire multiple litters of eight to ten piglets each year, while the females could only produce sixteen to twenty offspring.

In the case of the American Guinea Hog Association, there were only twelve foundation hogs documented in 2009. Half of these were male. The foundation hogs were purchased from 4 breeders. The breeders and foundation hogs are listed in Table One. In 2011, eleven hogs were identified as foundation hogs in the AGHA Newsletter, "Guinea Hog News," but offspring of Celesky's Lola were registered after that date, so I included her on Table One and identified full siblings. I will introduce the original AGHA foundation hogs, their breeders, and their owners so you can learn more about them from original sources.

Table 1
Foundation Hogs of the Early AGHA Registry Seeded the Gene Pool from 2002-2009 (Twelve Hogs, Three Family Groups)
1) J. Frank Baylis, breeder in Virginia Herd Source: Alabama breeder *Foundation Boar*: **Baylis VA Samson**
2) Bill Biggers, breeder in Virginia Herd Source: unverified *Foundation Boar*: **Biggers Arthur**
3) Randy Setty, breeder in Ohio and Mark Celesky, breeder in Nebraska Herd Source: unknown Mark purchased hogs from Randy and possibly one from Donna Watkins in Illinois *Foundation Boars*: **Celesky's Boris** **Setty Houdini**** **Setty's MC Little Old Stiff Guy*** **Setty's MC Big Old Stiff Boar*** *Foundation Sows*: **Celesky's Lola** **Celesky's Roxanne** **Celesky's Tulip** **Setty Lilly** **Setty Rose**** **Setty's MC Wart Side Sow*** *Littermates born 4/15/2002 **Full siblings

Part V
Foundation Stock and their Breeders in the AGHA

Chapter Six
Bill Biggers

"They had great personalities. One thing nice is that they were easy to move around and even when the boars had tusks, they were still easy to move around and gentle. They would go all over the place."
Bill Biggers

When I telephoned Bill Biggers in August of 2017, I told him that I was writing a history of the American Guinea Hog.

"The Forest Pig?" he asked. I had to remember that this was many years after he sold Arthur. The name American Guinea Hog had not yet been conceived when he sold Arthur, and Guinea Hogs had many other names in the 1990s, none of which included the word *American*. I found Bill after reaching several dead ends looking for him. Earlier that day, my friend, Matt Hunker, had emailed me an article from the *Daily Press*. It was first published in The *Roanoke Times* on May 2, 2005. The article, written by Pamela J. Podger, was entitled "A Morsel of History: Rare Hog Charms Park Visitors." The story featured Louise, a very friendly "Guinea Forest Hog" and Louise's owner, Bill Biggers. It documented her time as a "hog ambassador" at events in historic Belle Grove Plantation in 1998 and in Pamplin Historical Park in 2003.

Finding that gem of an article helped me identify Bill Biggers' location and his age. That, in turn, helped me find his contact information online. After a little sleuthing, we became friends on Facebook. Soon, I was speaking with him on the phone. Today's technology is truly amazing.

William Biggers was born in 1941, and he is an avid caver, or spelunker. His beloved sow Louise was gored by a Devon bull and died in 2007, he lamented. He got Guinea Hogs Arthur, Louise and Gladys "from a fellow over in Maryland close to St. Mary's around 1998" he told me on the phone. He got his first breeding pair, Punch and Judy, in Stanton, Ohio in 1997. Bill admitted that he did not have a good memory for details. "I don't keep records really good, and I'm sorry about that. But I can tell you everything I remember," he said. Bill sold his Guinea Hogs as far away as the Buffalo Zoo and Bangor, Maine. However, he was not the breeder of Biggers Arthur; he was the owner.

In my research, I try to verify accuracy of stories whenever possible. Memories can be tricky when you are recreating histories from years past. However, each memory may hold some elements of truth, even as it fails us. Mr. Biggers sold Arthur to Don Oberdorfer around 2004. Don recalled meeting Bill Biggers and the boar Arthur. "When I got Arthur from Bill Biggers," Don remembered, "he had two boars. One had a little white on the leg and he had Arthur that I bought. And Bill said to me, 'Well, do you want a pure Guinea or one with something else in it?' And I said, 'Well, if I am going to get started in the breeding, I'd better have a pure one.' And I know a lot of times Guineas are so inbred that they throw a little bit of white in them, even if they are pure, so that may not have been known about that pig."

Don continued, "We didn't know enough back then to pay attention to the folks that did have them. I do remember asking Bill where he got Arthur from. He said he didn't remember. And then his wife walked outside and said, 'Oh, you're buying Arthur. We got him at the Roger Williams Park Zoo in Rhode Island. I remember carrying him home in my lap.' And I'm pretty sure that he knew that. But, you know, he was not lying. It's just that nobody paid that much attention or cared that much."

So eventually, I dug a little deeper into the mystery to see if I could resolve the discrepancy between the memories. I eventually tracked down Tim French, deputy director of animal operations at the Roger Williams Park Zoo. Tim replied to me with the following email:

"Dr. Payne, your request for information about Guinea Hogs that may have been sent out from Roger Williams Park Zoo was forwarded to me. I checked our records going back to 1970. We have only had four Guinea Hogs in that time. A pair purchased in 1992, and two females acquired from Beardsley Park Zoo in Connecticut in 2009. The 1992 animals died in 2000 and 2009. We still have one of the females acquired in 2009. The other died in 2016. There have never been any Guinea Hog births at the zoo. The couple you referred to in your inquiry must have gotten their hogs from someplace else."

When I followed up with Bill Biggers via Facebook messenger, he stated that he got Arthur at the Roger Williams Park Zoo. Zoos are known for their meticulous record keeping. Perhaps the hog did come from someone in Maryland as he originally stated to me. The origin is, unfortunately, not definitively known.

Apparently, Mr. Bigger's sow Louise was popular with the press and with local parks. There was an article written about her when she visited Sand Point Park, as well. "We were down there," Bill said, "and that is when she got the reputation for loving music. The banjo picker was tuning up his banjo in the morning. He went down there and started picking, and Louise started singing with him. She was keeping time right along with him."

Bill loved the Guinea Hogs, or Forest Hogs, because "they had great personalities and were easy to move around. Even when the boars had tusks, they were still easy to move around, and gentle." He said that, at one time, he had fifty-two of them, descended from the original five hogs. One of the reasons Bill got the hogs originally was due to their reputation for grazing and not rooting. He kept them on seventeen acres.

The group of cavers Bill belongs to have acreage in Deerfield, Virginia, where they hold festivals. Bill liked to bring whole hogs to these events for everyone to enjoy as barbeque. Bill wanted to share more stories about Louise, who must have been his favorite sow.

"The African Guinea Hogs are very laid back. When Louise was having piglets, she would let me come in there and touch them almost right after they were born. I mean you have to watch them, she might come after you. But normally they won't. She would know that you were taking care of them and you are like their pack and their pack leader. The worst thing they did was get out underneath the fence and they would go over and tear up a guy's flower bed. I replaced a lot of flowers."

Bill wasn't finished yet. "Louise was down in Rowan Oak Park and they had a school group come in. And the guy told them to stay back away from her mud hole, which was right next to the fence. He watched them, but nobody believed him and they all came up next to the fence. She had piglets with her because she gave birth at the park. The guy saw her look over and the next thing she does is trot straight to the mud hole and flop herself into that mud. It just took up a geyser of water that went right over the fence. Got every one of them school kids. Then she rolled a couple of times and then got out of the water hole and went back to her piglets. She knew what she was doing."

Guinea Hogs have a variety of coat types, from soft to stiff bristles, sparse to thick coats, and straight to wavy or curly hair. It had been my theory for a while, looking at pedigrees, that the shared foundation animal carrying the trait for curly hair was Biggers Arthur. In my herd, I had a curly-haired sow whose parents were great great grandchildren of Biggers Arthur. Each of them had this boar in their pedigrees four times. My experience with my sow LSF Bess was that in each litter, some would develop curly or wavy hair by age eight months old and some would have straight hair. All of them had very thick, long hair and a large, rounded frame. Noses were on the short side. They leaned toward the big-boned type. The boars were docile and loved belly rubs. My interview with Bill confirmed that his herd carried similar traits.

Don Oberdorfer told me that Biggers Arthur was more "feral looking" than the Celesky and Setty hogs. He had large shoulders and narrow hips. Don described it as "almost like a cartoon bulldog." He also stated that Arthur and his son, DNC Junior, were independent and liked to be scratched. Arthur was also a bit *snufflier*. Sort of like a bulldog, but he would do that in a good way when you scratched his ears. He had more personality than Houdini."

Pedigrees that include VAZ John Henry and VAZ Harriet will be linebred to carry traits of Biggers Arthur. I linebred my LSF Bess to one of her sons in order to bring Arthur's genetics up to twenty-five percent of the genetics in the offspring I selected from that litter. That boar is BRP Sonny Boy Williamson, the cover boar for my book. He has quite the curly coat and is full of personality. Shannon Engelhardt is Sonny Boy's current owner.

Chapter Seven
J. Frank Baylis

"But when there was a sow in heat, that pig could fly!"
Shirley Sullivan

"My hogs had short legs and a lot of body. I didn't want them
too long in the leg. Nice curly tail, very hairy, short muscled.
Not a narrow head, but a v-shaped head. It was a nicely shaped head.
Very powerful pigs. Small, but powerful.
Extremely hairy with a good head."
J. Frank Baylis

Baylis VA Samson

My very first boar, and the only one I had for my first two-and-a half-years of breeding hogs, was Sullbar VA ML Porgy, a Baylis line hog. I learned about the Baylis line from Shirley Sullivan. She contacted me when she had a large litter of piglets that included a promising boar. I didn't have the time to drive to New Hampshire, so Shirley offered to ship him to me in a crate via Delta Airlines Freight. That was such a generous offer that I took her up on it, and Sullbar VA ML Porgy came to Georgia. The "VA" after the farm prefix indicated that Porgy was from the "Virginia" line of hogs, although Biggers Arthur was the first hog referred to as a Virginia line and unrelated to the Baylis hogs. The "ML" came from the first letter of his sire and the first letter of his dam to form a "cheat sheet" of his breeding. I borrowed this idea of using a name prefix to denote specific lineage when I later acquired the Sumrall hogs.

Once I got settled in with my new herd, I had the opportunity to interview Shirley, and later her husband Jim Barnett, for my book. She explained to me they had wanted pigs for the ranch, but she was nervous about having huge boars on the property. Jim had read the article Don Oberdorfer wrote recommending Guinea Hogs. They spent the next couple of years acquiring stock from Don Oberdorfer, Kevin Fall, and Callene Rapp, keeper of the Children's Farm at the Sedgwick County Zoo. One day, Jim was searching for other zoos with Guinea Hogs and located the Morris Farm in Wisconsin. It is an educational farm for children. They had an elderly boar named Samson.

Samson and his mate, Delilah, were donated to the Morris Farm by a couple in Maine. They didn't have time for them and thought it would be better if the pigs were at the Morris Farm. This was in the 1990s. Samson was born in 1994. Delilah died during farrowing, leaving Samson without a mate for the next decade. The couple who originally owned Samson told Jim that they had purchased the duo from Frank Baylis, a dog show judge in Virginia.

By age thirteen, Baylis VA Samson was an elderly, obese, arthritic boar. The Morris Farm administrator was not ready to give him up and would not allow Shirley and Jim to loan them a sow. The administrator did not want children witnessing the breeding process at the farm. So in the summer season, Jim and Shirley brought Samson to their New Hampshire property, Sullbar Farm. They did this for two years in a row, making a fourteen-hour trip each time. At Sullbar Farm with its open pastures and sows in season, Samson found a new life and enjoyed his excitement-filled vacations. After the second summer away from Morris Farm, the staff were unable to contain him in his pen any longer. "He realized that life was a lot nicer if you can move around," Shirley said. "And they called us and said, 'You come and get him and take him home.' "So we did." Shirley described Samson's condition for me.

"He was so crippled when we brought him down that he couldn't walk. His hind end just swayed. So his front feet would move, and his back legs would almost drag. He pretty much never recovered from that. But when there was a sow in heat, that pig could fly! And that is a true statement. We let him out of the trailer to put him in with a sow. And that pig, he had not bred for I don't know how many years, he jumped right on the sow. Ran right over and jumped on the sow! I was shocked!" she exclaimed.

Shirley told me about elderly Samson and his progeny. "The personality is completely docile in most of his offspring. Samson and his sons and grandsons love to be petted. Udder Hope Matt [Samson's son] will ignore his food if he thinks I am going to pet him. There is a definitely bright brown eye [in Samson]. Some are the darker eyes, but his were bright brown and crystal clear, right up to the end. Any of the offspring boars we kept looked like him—very flat shouldered and square legged. Now Samson himself, he was so crippled, you hardly saw all that. His hind end wound up being much lower. His hips were a mess. We never got to see the fullness, how strong his hind end would have been in his younger years. I will send a picture of him when we first got him. He was in poop up to his knees. It was cold, it was wet, and he was at that children's farm. I didn't even recognize him as a Guinea Hog because he was such a mess. It was just horrible. I have to say that because of his physical problems, unless testosterone took over, he didn't move around a whole lot."

Shirley continued, "You know, he walked from his house to his water and to his bed. He laid down a lot. I am sure it was painful for him. But when the testosterone was flowing, he could climb things that you would not think any pig could climb. Like a four-foot stone wall—he would go right over it! He would get in physical fights with young boars. They would fight shoulder to shoulder, and that would be it. He would be exhausted. I came home more than once and thought he was almost dead because he had broken out and broken into a younger boar's pen and gotten into a horrible fight with him. He was lying in the sun, panting, and I thought he was dying. I gave him some water, and he got up and went back to his pen."

Shirley and Jim made the difficult decision to put Samson down when he was just short of eighteen years old. He had a swollen testicle, and the vet thought that it was probably cancer. He got to the point that he could not lift his head to eat. I am grateful to Shirley and Jim for putting in the time and effort it took to build a relationship with Morris Farm and bring such an old specimen of Guinea Hog home to continue a genetic line. I discovered in my interview with Mr. Baylis that Samson was the only remaining representative of a very old Alabama lineage.

Since the twelve hogs that formed the foundation of the AGHA registry were from only three or four lines, Samson represented one-fourth to one-third of the genetic family strains in the national herd. In addition to that, his DNA was 75 percent unique compared to Biggers Arthur and the Setty family group. If Sullbar Ranch had not taken on this project, it would have been a great loss to the breed. Jim and Shirley contributed to the breed in other ways, as well. Shirley served as registrar to the AGHA registry for several years, and Jim served as president of the AGHA.

Foundation lines of the AGHA registered stock are traditionally referred to by the breeder's name or the name of the breeder's farm. For example, "Setty line" when the breeder's name is Setty. You may hear people refer to the "Samson line" of hogs, but there were no breeders or farms named *Samson*. To be consistent, the correct terminology is "Baylis line."

Shirley and Jim worked to develop a line of hogs from Baylis Samson, indicating related hogs by adding the "VA" name prefix. If you are interested in continuing that line and the deeply muscled look his descendants carry, a good start would be to look for pedigrees with lots of Sullbar Farm stock, then look further for the VA prefix or Baylis VA Samson in multiple generations. There are several breeders on the east side of the country with those genetics, some in the southeast, and at least one in Michigan. My former boar, Sumrall VA ML Porgy, is now at Ham Sweet Farm with Kate and Christian Spinillo.

J. Frank Baylis

Joseph Frank Baylis is the owner of Bayshore Kennel and Barn in Toms Brook, Virginia. His friends call him "J. Frank." To breeders of the AGHA, he is known as Frank Baylis, which is how the name was recorded in the herd book. Bayshore Kennel and Farm's motto is "Dedicated to preserving and protecting rare breeds of livestock and canines." It was natural that J. Frank would gravitate to the rare Guinea Hogs, and he first started breeding them in 1982. This was more than two decades prior to the formation of the AGHA. His original stock came from an eighty-one-year-old Alabama breeder who had raised them all his life, so it is possible that the herd had been closed for many years.

Other breeds J. Frank raised included Bluefaced Leicester sheep, long-haired fainting goats from Nova Scotia, and several rare dog breeds including Affenpinscher, Carpathian Shepherd, and Xoloitzcuintli. J. Frank loved the "minor" or heritage breeds for his livestock. He had heard that Guinea Hogs were not as bad about rooting as other hog breeds, and this appealed to him. They did not disappoint him on that point except after a rain, when they "rooted like crazy."

J. Frank started his herd with six sows and two boars from the Alabama herd. He got three sows from two different bloodlines (six total sows) and two boars from different backgrounds (two total boars), or four in a breeding group (three sows with one boar per group). Basically, he began his herd with two family groups that included sows bred to distantly related boars. Eventually his herd grew to five breeding boars and thirty breeding sows—six sows per boar. Because he kept a closed herd, J. Frank's hogs became inbred over the years.

He had a good meat business from culling offspring that did not meet his high breeding standards. "We sold them," he told me. "A friend of mine had a smoker, and we would smoke suckling pigs and do parties. I would supply the pigs, and he would supply the smoker. They were really good. Older hogs got too fatty, but the roasters were not fat—they were heavily muscled. It was just the perfect hog for that—not too big. Regular hogs get too big, too fast.

J. Frank described his stock selection criteria. "I had short-nosed Guinea Hogs. They had a really nice short face and were really hairy. My favorite boar was called Boarella. He was actually our last boar. His hair hung in ringlets. It was kind of wavy and curly in combination. He had the most hair and the best head compared to other boars. It is a hairy breed, but he was almost over-coated. He had a lot of hair. He was just the perfect combination, and, like I said, a beautiful face. Like a pig that has gone feral. Not like a Berkshire. But I selected for a short face."

J. Frank continued, "I'll tell you also what we got sometimes from being very inbred over the years was a white cornet band on the back foot. We got it from being terribly inbred over the years. I stopped buying outside pigs. I just inbred, and then I got rid of the pigs. White feet went to slaughter right away. You don't want that. Colors, whenever you breed animals, you can change and make new colors with purebred stock. It is like chickens. There are a million chicken breeds and they all stem from two breeds of chickens. And everything is descended from jungle fowl. So over the years, colors crop up, like in dogs. You can have a black breed of dogs and then all of a sudden, a red one. I was like that, but I got rid of that. It was the biggest problem I had when breeding Guinea Hogs. Some people had health problems, but I had the cornet band. One out of twenty-five pigs would have it. Only one foot."

He described other characteristics selected for in his herd. "My hogs had short legs and a lot of body. I didn't want them too long in the leg. Nice curly tail, very hairy, short muscled. Not a narrow head, but a v-shaped head. It was a nicely shaped head. Very powerful pigs. Small, but powerful. Extremely hairy with a good head."

Boarella eventually went to a woman in New Market, Virginia who wanted breeding stock to add to her goat farm. She also took three sows. J. Frank sold his stock due to their omnivorous habits, he told me. Back in the 1980s, before portable electric fencing systems were commonly available, it was harder to separate stock on pasture. J. Frank told me his astonishing story.

"I'll never forget the first indication of what was going on. I was doing dishes. We had a huge flock of Scottish Blackface and Bluefaced (Leicester) sheep. I could see them out of the window, and I saw one lambing. A pig walked up and sniffed the ewe as the head came out. In a blink of an eye, the pig grabbed it out of the ewe and ran with it. I went out and shot her because I knew that was the problem. I didn't have any more problems until I moved to the new farm. I had already downsized before the move to five sows and a boar. I heard a new(born) goat screaming. I went around and saw two sows fighting. One had a front leg, and the other had the other front leg. They were fighting and playing tug of war with that goat. I went and shot those two sows and called up the food bank. I put the other three sows in with the boar in a pen. That's the moment I decided to sell them, so they went to the lady in New Market."

That was around 1995. Anyone looking for old herds of Guinea Hogs might try to locate the woman in New Market, Virginia who bought the last of that old Alabama bloodline.

Chapter Eight
The Setty Line Hogs: Setty, Celesky, and Watkins Connection

"Until we formed the AGHA, there were few people who kept any kinds of records. At that time, people really didn't think it mattered all that much."
Don Oberdorfer

Mark Celesky

Kevin Fall warned me that it is hard to get in touch with Mark Celesky in the autumn months, and he was right about that. Mark's business in Cerasco, Nebraska is Boarding House Farm, but he works with Vala's Pumpkin Patch in Gretna each year during the pumpkin harvest. Vala's is a huge venue for agritourism, with everything from hayrides to corn mazes to pig races, and open seven days a week from mid-September to Halloween. Mark has participated in over 8,000 shows over the years.

Kevin told me about the stock from Mark Celesky that ended up in the AGHA registry. "Genetically they were tested, and there was one group that were possibly all siblings, and another group that wasn't."

SAVING THE GUINEA HOGS

Mark was very forthcoming during our interview when I finally got him to pick up the phone. When I told him the hogs he bred were one of a handful of foundation stock for the AGHA, he laughed a bit and said he was relieved for a chance to set the record straight. A friend had shown him registration papers with his name on it, and that had surprised him, since he had never been part of the AGHA and did not keep a herd as such.

The story, from Mark's recollection, is that he purchased about five hogs from Randy Setty. He kept the Setty hogs and also borrowed boars to display on the farm from 1997 to 2005 before the AGHA was established. He was not a conservationist, but had plans for breeding that "never fully materialized." He explained, "I only bred two litters of Guinea Hogs." Two sources told me he purchased at least one hog from Illinois, which would likely have been Donna Watkins' stock. When I asked Mark about that, he said the name sounded familiar. However, Kevin recalled more clearly that Mark had obtained "Illinois" (Watkins) stock and coordinated taking them to the Sedgwick County Zoo. The zoo has had Watkins hogs in their breeding records. It is not surprising that Mark did not recall where he traveled twenty-three years earlier to get some hogs.

Mark wanted piglets on the farm each October, so he traded with people like Randy Setty and Kevin Fall. Basically, he would borrow some boars, keep them on display through the pig races and pumpkin patch events, and then sell the shoats to interested parties. Mark confirmed sales from his two litters and re-sales of borrowed hogs to Paul Krumm, Dodge Nature Center, and Sedgwick County Zoo (SCZ). The SCZ purchased hogs in 2000, 2001, and 2005. Because of the time gap, it is likely that the stock taken there in 2005 would be some he had borrowed, and not bred from his Setty stock. The 2005 hogs could have been obtained from Donna Watkins, Paul Krumm, or Kevin Fall. These were all people that he traded with. Sedgwick

County Zoo records show that he purchased a litter of hogs in 2005. Perhaps these were used for additional trading.

Mark sold or traded three of the Setty hogs to Kevin Fall. Kevin added the initials MC to their names to designate who he had purchased them from, but kept the breeder name. Those hogs, littermates born in April 2002, were Setty's MC Big Old Stiff Boar, Setty's MC Little Old Stiff Guy, and Setty's MC Wart Side Sow.

Randy Setty's Herd

As far as I could determine from public records, Randy still lives in Goshen, Ohio outside of Cincinnati with his wife, Bonnie Jean. A letter sent to an old PO box came back with "address unknown" stamped on it, but two letters to his Goshen home address were never answered or returned. The phone number I located rang, and I left voicemail messages, but I never made contact with Randy. This was a real blow to my research, since ten of the foundation hogs came from Setty's farm. I would love to know the origin story of his herd, how closely he linebred, how many breeding sows he kept, and if he had more than one boar. If he had only one breeding boar, all of the Setty hogs would be at least half siblings, with the Celesky hogs closely related.

Kevin Fall shared the following information with me. "We now know that the Setty and Celesky hogs were actually related genetically. And that Rose and Houdini were full siblings. Hindsight is 100 percent. We didn't know then what we know now." Kevin told me Celesky Tulip got too fat and stopped having piglets. The hogs in the Setty herd tended to grow quickly and put on weight readily.

Kevin assumed that Randy had passed away due to a serious illness he had in his thirties, at the time when he was selling his stock. However, I did not find an obituary when I searched his name. I reached out to several Virginia people named Setty on Facebook, but they all denied having a relative with that name.

John Ross, Jr., of Foggy Bottoms Farm in Cynthiana Kentucky, purchased Setty MC Little Old Stiff Guy from Kevin Fall. The boar is now deceased. I think the long story is that first he purchased Setty Blackjack Luther, but Luther was not interested in breeding. Kevin swapped him out for Stiff Guy, who did fine. Luther went back to Kevin but was hit by lightening and killed.

John told me what he liked most about the breed. "I liked the Guineas because of basically how durable the piggies are. They are up and running around in less than an hour, so they tend to not be squashed. The last litter I had was starting to eat solid food within four days. I like that about the piggies, and the sows are generally good-natured. When they have their piggies, they are in a stall. So I have to go in there at least twice a day to feed and water them. It is important to have a sow you can deal with."

John also shared his observations about Randy Setty, based on his visit to the Setty property in 2003. "Randy's place was just a mess. Confusion, and he also had some other small breed hogs there. He was in terrible shape when I was there. He was using two canes, and he could barely walk. And that is why he was getting out of hogs. He told me he was afraid that he would fall down in there and get eaten. In fact, he wouldn't get into the pen unless his wife came out with him.

"The breeding pair that I got from him just did not work out well at all. I never got any piggies from the sow. He sold the sow to me as bred. It was believable because she looked pregnant. He told me he didn't feed them regular feed. There was a Pepperidge Farm place down the road, and he got all of their old baked stuff, so that was all they ate — bread, cookies, and donuts. The poor old boar was so fat he must have weighed 400 pounds. He dragged bottom, and the sow was not much better. They never produced anything and then they both died. I paid about 700 bucks for the pair of them."

John and I chatted about some of the sows he owned over the years. He recalled Sugars Shadey Adie. "It is funny," he said. "She was very different in appearance than the other ones. She had a really long snout. She looked like an anteater. She was also a bit longer in the back than the others were. A lot of her offspring looked like that as well."

(Author's note— Sugars Shadey Adie was bred by Jessica Benson from a DNC duo. Her sire was out of Houdini and Rose, and her dam was out of Biggers Arthur and a great granddaughter of Biggers Arthur. Those surprisingly long noses can pop up out of nowhere, it seems, and then begin expressing in the next generation)

I asked John if his hogs ever threw piglets with white markings.

"Yes, in every litter," he replied. "There would be one piggy with white toes, and half the litter would have a couple of white toes or a couple or three white socks. And I have been told that it is something the Setty hogs have. "Setty and Celesky stock will produce offspring with white noses and white feet on occasion or even frequently, judging from John's experience. If parents expressing white are bred together, the offspring may express what the AGHA standards refer to as "excessive white." Don Oberdorfer told me that in "Setty offspring I see a lot of a white spot on the nose and a red flash in the mane."

The Setty MC offspring and great-offspring were docile and grew large on milk and pasture. A little grain went a long way. Kevin told me, "The reason I got into the Guinea Hog is because they tested them in different places and different people tested them by putting them out in a lot with grass. They got almost too fat to walk just from what they could forage themselves. They would find acorns, earthworms, and graze on the grass. They have noses like a bloodhound and can sniff out anything growing in or under the ground. They hardly needed any grain. We were well pleased with the way they performed, and that was the push point years ago."

Setty Lilly

Setty Lilly was one of the first hogs owned by Shirley Sullivan and Jim Barnett of Sullbar Farm. They got her from Kevin Fall, who picked her up from Randy Setty. She was one of the founder sows. Shirley told me that she was an amazing pig and had several healthy litters before she died during farrowing. She had been an excellent mother. Jim told me that Lilly was his favorite sow. Shirley describes her as very square and thick-boned with medium legs. Her hair was blue black. "She was an exceptional mother, very healthy. She ate well, stayed in good condition, and always bounced right back after a litter."

Setty Rose and Setty Houdini

When I interviewed Don Oberdorfer the first time, I asked him about Setty Rose and Setty Houdini, the full sibling duo. He got them from Randy Setty in Ohio when his cousin got married in September 2002, as described earlier. Here is a more complete story Don told me about that trip back to St. Paul.

"So when we went to my cousin's wedding in Ohio, I met Randy Setty. We swapped out the pigs and were coming back from Akron along the Ohio Turnpike. We were in Indiana, and it was one of those really brutal summers where it was over 100 degrees. I had just put the pigs in the back of a pickup truck with a topper on it. So we pulled into the truck stop and we were just eating some fast food. And both Houdini and Rose at that point were way overweight. They were really heavy, heavy pigs to the point that it was not a good thing. So we were sitting in the McDonald's, and my brother-in-law comes in and he says, 'The pigs are loose, and they are running all over the parking lot!'

So we went out there, and those hogs were heavy and hot! It was not super hard to get them to go back towards the truck, but we had to actually go arm in arm underneath them and lift them and try to get them back into the truck. Then I had to go into the convenience store and buy bags of ice and pack them down so that they wouldn't be overheated. And there is a guy from out east somewhere who had like a little convertible sportster and he was just standing there and saying, 'I must be in the Midwest.' Like he thought pigs run around parking lots all the time there."

At the time of our interview, Setty Rose was still alive at Dodge Nature Center, so she is described below in the present tense. She died in 2015 at age fifteen years. She became infertile after age ten years. Houdini died in 2013, at age thirteen years. Don described his Setty hogs as having short, squat legs compared to Biggers Arthur. "Houdini was very rectangular and his hips were about as wide as his shoulders. His hair was really thick, and he didn't shed that much. Rose actually shed a fair bit. They both had the eyes that looked forward at you. I find Rose to be really pleasant, but a little bit bigger. She is a little more dominating to some of the other pigs, but maybe that is just being an old lady and not wanting to be bothered. Both Rose and Houdini had the propensity to get fatter if overfed. They would bulk up way quicker than the other ones would, so I had to be more careful on how to feed them and cut their weight if they were getting too heavy. If they get too heavy, their fertility goes way down.

"Compared to Biggers Arthur and his son, Junior, Houdini was a bit dumber and lazier. He was a friendly animal, but there was less personality to him. When I got Houdini from Randy he was a mature boar with tusks. My brother in law and I just picked him up, and he didn't know us at all. And he just let us do it. That is the good natured side and part of the reason we like Guineas. Rose is a curmudgeon, but very nice."

Setty's April 2002 Litter

Three of the Setty foundation hogs were littermates—full siblings from the same litter. They were Setty's MC Wart Side Sow, Setty's MC Big Old Stiff Boar, and Setty's MC Little Old Stiff Guy. Mark Celesky obtained these hogs from Randy Setty and then sold them to Kevin Fall.

Setty's MC Wart Side Sow is owned by Kevin Fall. She was alive and twelve years old when he told me about her in 2014. She had a "litter plus one" before she lost her fertility. Her first litter for Kevin was born in April 2008 when she was six years old. Four piglets — two male, and two female—were selected from that litter. Her next farrowing produced only a single piglet. After that, she did not produce again.

Kevin also owned Wart Side Sow's sibling Setty's MC Big Old Stiff Boar. Kevin told me "I got some of his genetics out there." One of the boar's sons was sold to a breeder in California. Kevin had planned to do some linebreeding to "recreate" Stiff Boar but first the boar's daughter died, and then the boar died. Kevin described what he liked about Stiff Boar.

"He was probably more so of a perfect offspring and some of his offspring were the Ohio line that I think are really ideal. Really erect ears. Nice look in that aspect. But they are also all around they have body. They are not just shoulders, and they're not just hams. They are built all the way around. So they have good build front and back... I tell people that you want a Guinea Hog that when you look at them, their backs are flat and they look like a rectangular shaped box. So you have flat sides, you know, you will have some roundness and a head on that, but you want a body that looks like a rectangle. Like a box, not like a barrel with pigs underneath of it. That is the biggest problem that a lot of people make is that they get them fat and then say, 'Oh, now what am I going to do?'"

He continued, "His hair was medium long. He was not long and some of these are really curly, and the really long haired, he was not that way, but medium long haired. Like a nice Guinea Hog."

The third littermate born in 2002 was Setty's MC Little Old Stiff Guy. Kevin traded him to John Ross Jr. when he had trouble getting Setty Blackjack Luther to breed. Later, Luther was hit by lightning and died. I am going to assume that Stiff Guy did not have the erect ears that Kevin admired in his brother Stiff Boar. That is because some offspring and some great-offspring of Stiff Guy have very forward facing ears—more than I have seen in other Guinea Hogs. These descendants are big, bulky hogs with gentle personalities, a tendency to flop, and a proclivity to fatten on grass.

John Ross raised the hogs primarily for his own meat and enjoyment, not to produce breeding stock. Three people who purchased hogs from him were Tammy Redmond, Fran Collins, and Jocelyn Stanfield. Jocelyn was the last to get a breeding duo—FBF Sycamore Clyde and FBF Sycamore Bonnie. They are half-siblings with Setty's MC Little Old Stiff Guy as sire to each. Jocelyn dispersed offspring from two or three litters of these hogs to several farms in the eastern side of the country, so these genetics are available. Alesha Gonzales in the Augusta area purchased the sow Bonnie after she was bred to Clyde one last time. I purchased a boar, UGF BB King, from that litter. He is now owned by Sandee House of Friday Farms. BB is one of those boars who sees you coming and calmly lays down to offer his belly. He loves a good belly rub!

The offspring of Bonnie and Clyde have Sugars Shadey Adie as a grandmother. She is the sow with a "nose like an anteater." While I have never seen that characteristic pop up, it is not unlikely that it could skip a generation or two and then make a surprise appearance.

Celesky Hogs at Sedgewick County Zoo

Callene Rapp, senior zookeeper at Sedgwick County Zoo in Kansas, has been a caretaker of three of the four Celesky hogs foundation hogs—-Celesky's Boris, Celesky's Roxanne, and Celesky's Lola. Don Oberdorfer raised Celesky's Tulip in St. Paul. Lola is alive and well, turning fourteen years old in 2019. Boris sired Lola's last litter before he died, producing SCZ Kit and SCZ Kat. These littermates are now seven years old and live at the zoo as exhibit animals. Lola's genes live on through SCZ Fred, an offspring owned by Jack Rowland in New York State. She also produced the boars SCZ Hampton and SCZ Boaris. These boars may still be breeding. SCZ Hampton sired a blue hog that remains at the Beardsley Zoo. There is no boar currently living at the Sedgwick County Zoo, and Kit and Kat are not likely to ever breed. Celesky's Boris lived to age eleven, and Celesky's Roxanne reached thirteen years old before passing away in 2013.

Callene told me that the hogs she has cared for are predominantly solid black, although some have had minor white around a single toe or dewclaw. She told me that Boris and Lola had the rectangular shape that Kevin Fall described in the Setty hogs. Lola also had a long snout. Longer than typical, she explained, based on limited experience with them in the zoo. None of the Celesky hogs kept at the zoo ever turned blue, and none developed curly or wavy coats.

The Sedgwick County Zoo appreciates the gentle nature of the Guinea Hogs, particularly that the boars have never posed any threat. They have been easy to manage due to their size and dispositions. Callene told me, "The thing that I like about them is that they've never given anybody any management issues. Even the least experienced person can get along with them reasonably well. They are small in size and easier on the exhibits. They are just easy to manage."

The zoo had some weights in the files recorded prior to 2016. Lola was 163 pounds, Kat was 148 pounds, and Kit was 134 pounds. Callene stated that they were in good condition with a "nice kind of round tube-shaped hog look." She described the hair coats as straight, thick, and long. Sometimes the hairs get split ends and bleach at the end.

Close Relations in the Setty Line

Don and I discussed the DNA testing that confirmed that Setty Rose and Setty Houdini were full siblings. This knowledge came after Rose and Houdini had mated and supplied numerous offspring to the small population of Guinea Hogs available for registration when the AGHA began. When you start with twelve hogs from limited family lines, you have to start with what you have. The initial push was to get numbers up. Don shared his thoughts about inbreeding and selection in the breed.

"The truth is that the Guinea Hogs are really inbred. We need folks that breed them to be fairly conscientious breeders. Now we are not worried about how many we have. We are worried about health. Your pigs, do they have deformities? Do they have problems that are noticeable? Are you culling stock or breeding everything? Are you subtracting the ones that should be subtracted out because they are adding stuff to the gene pool you don't want from an inbred population? That is what matters more than numbers. Then they can last a long time inbred. This one has bad legs, and it needs to be delicious. This one has hip problems, and it needs to be delicious. If you ever get one that's mean, it needs to be delicious."

Some breeders with high inbreeding coefficients that include Setty Rose and Setty Houdini should be aware of an issue related to the software that computes that percentage. An inbreeding coefficient tells how high the level of inbreeding is based on ancestors who are repeated in the breeding. Because the founder hogs' relations were unknown when the registry started, their parents and grandparents are not computed into the formula. From my limited experience with the formula, my understanding is that the numbers have less meaning in the first three or four generations of initial calculations in a breed because the data is missing. Don explained to me, "We had to start somewhere." It wasn't for a few years that DNA samples revealed that Setty Houdini and Setty Rose were full siblings. Randy Setty did not keep records and did not share how related they were. Nobody knew to ask. Now that we know, those working to continue with Setty linebreeding need to proceed with caution to avoid running into low fertility issues or health issues.

As an example, I found registration papers from a Setty linebreeding. The boar was a son of Setty Rose and Setty Houdini. The sow was a daughter and a granddaughter of Setty Houdini. The Coefficient of Inbreeding (COI) on the pedigree was listed at 25 percent, which is quite high, due to the repetition of Setty Houdini in the pedigree. I put the data into a pedigree software program and made one change—I fabricated a name for Houdini and Rose's dam and sire and added that information where there were blank slots. By adding their parents, who now also repeated in the pedigree, the COI jumped up to 43.75 percent. That is extremely high, but more accurate than the original number. I imagine the same thing would happen with the FBF Sycamore Clyde and FBF Sycamore Bonnie offspring. They are listed as 16 percent COI but that is probably closer to 22 percent.

That makes them ideal for linecrossing, but not for linebreeding more than one generation. Anyone using hogs such as these in a linebreeding program would be wise to linecross in alternate generations to avoid reduction in vigor and fertility. Each generation should be tracked for growth rate, overall vigor, and reproductive success. Celesky hogs should not be used for linecrossing with Setty, as they are essentially Setty line hogs, too. I hope that some breeders continue to isolate the Setty lines and breed them, but to use wise strategies and linecross in alternate generations.

Part VI
Discovering the Missing Genetics

Chapter Nine
Breeders of the Lost Genetics

"Everybody back in the day would say, 'There are five family lines.' One was Annette Hesters, one was the Sumralls in Mississippi. We had Setty in Ohio, Biggers in Virginia, and Nebraska was Mark Celesky....If I found these family hogs and was on the board [of directors of the AGHA] today, I would probably include them. Like Phil says in his book about preserving breeds, one of the things that he talks about is you want to be as inclusive with a narrow gene pool as you can be."
Don Oberdorfer

"In the case of the Guinea Hog, it will be especially useful to track down any source herds that have been overlooked in the past—these will greatly help to provide the genetic strength that the breed will need to move into the future.
D. Phillip Sponenberg

Introduction

There are only a few known breeders who maintained Guinea Hog herds that were established prior to the formation of the AGHA in 2006 and were still actively breeding at that time. This section introduces some of those known breeders. After purchasing the Setty/Celesky/

Biggers herd from Stephen and Hollie Brothers in 2005, Arie McFarlen of Maveric Heritage Ranch obtained stock from Billy Frank Brown, Annette Hesters, and Gary Sumrall. These transactions occurred before the AGHA was formed, and the sellers involved reported details to me. Annette Hesters had purchased her original hogs from Dan and Shirley Hale, Marcia Read, and Donna Watkins decades before the AGHA was established.

Thankfully, the genetics from these bloodlines were retained and preserved by McFarlen, Sumrall, and Hesters. However, none of these breeders joined the AGHA when it formed after their herds were established. Today, the majority of Guinea Hog breeders own and breed *unregistered* Guinea Hogs, and never join the breed association. Those genetics are therefore undocumented and do not contribute to the future gene pool. Without records, the inbreeding coefficients are unknown, and the bloodlines can hit a genetic bottleneck. When that happens, breeders will wonder why their hogs are infertile, have diminished litter size, or are not thrifty on pasture. In this chapter I introduce readers to those known breeders who were actively breeding before the AGHA was formed. Of these, only Annette Hesters is still breeding in 2019.

Billy Frank Brown and Fred Keene

I was truly fortunate to track down Billy Frank Brown and eventually meet his son, Jess, who is now running Cowpen Creek Farm in Poplarville, Mississippi. Jess is active in The Livestock Conservancy and is continuing his father's excellent recovery work with Guif Coast Native sheep, Pine Tacky horses, and Pineywoods cattle.

The Brown family maintains some of the most unique landrace livestock in the United States, according to their farm brochure. These species descend from Criollo stock brought over by explorers as early as the 1500s. The family homestead known as the "Sebron Ladner Place," was first settled by Billy Frank's grandfather in 1811. The family heritage traces back to 1726, when Nicholas Christian Ladner married Marie Anne Paquet, a Native American descendent of the Gulf Coast. Billy Frank is a ninth-generation farmer, and Jess is the tenth-generation. Billy Frank's twelfth great-grandpa moved to the Mississippi coast in 1699 and is credited with bringing the Pineywoods cattle into Biloxi from France, Mr. Brown explained.

I've had three interviews with Mr. Brown. The first was in December 2015, the second in June 2016, and the third in July 2017. Mr. Billy Frank Brown is a real gentleman; generous about sharing his knowledge and memories. Billy Frank was born in 1942, when times were different. All of Cowpen Creek's livestock roamed the open range, as did their neighbors' stock, and many other herds throughout the Southeast and the Gulf Coast.

When Billy Frank's grandfather, Sebron Landner, passed away in 1956, he had one thousand head of Pineywoods cattle, four thousand Gulf Coast Native sheep, and forty Pine Tacky horses roaming on open range. The hogs ran the same way, and Billy Frank remembers that his grandpa Landner had Guinea Hogs.

"I remember when I was twelve, fourteen years old, you know. We had them. That would have been in the 1950s, you know. They were, you know, it's not really a stock of hog. It's the old-timer hog line, you know." Like the Pineywood cattle, the Pine Tacky, and the Gulf Coast Native sheep, the Guinea Hog is a landrace breed, formed by local pressures to adapt and with distinct family strains. Billy Frank explained, "They'd just keep breeding them back. Now I've just got some Pineywoods hogs, not any Guinea stock. The last Guinea Hog I had was a boar I sent to South Dakota. Let's see, her last name was McFarlen. She came by and picked it up. It was a blue Guinea boar."

D. Phillip Sponenberg had told me about an elder veterinarian in Mississippi who claimed that the blue Guineas once roamed widely in the state. I asked Billy Frank to tell me more about his blue boar.

"Well, way back, you know, we had blue hogs wild in the woods. There are still some of the strand of the blue hogs in the woods still, especially in Georgia, I think. Not so much as there is in Mississippi. But in Georgia they still have the blue colored hogs. The one I gave to Arie weighed about, I'd say, 150 pounds. I fed him good, you know? He's a Guinea because of his short legs. Pretty long body, but he's short legged.

"I raised him from a sow here. He was raised here on the farm. He was probably a three-year-old when she got him. He was a pretty good size. You know, a pretty old pig. His daddy was blue, I think, if I remember right. They were out of some of the old stock that came out of the Mississippi woods. He just turned out to be a Guinea, and I just kept him for a boar. I had a black boar, too, what we used to call the Andrew boar. Arie came along. I wasn't really trying to raise Guineas. She took him to South Dakota."

He continued, "The first old-timers out of South Carolina and Georgia, when they came in the 1800s, they brought them in here and turned them loose. That's what they raised their family on. They brought them with the cattle, horses, and sheep in the early 1800s about 1806 and the early years. My daddy and my granddad had Guinea Hogs in the woods, too, and probably my great-grandfather. We've been here since 1811. My family's been right here since 1811. Carlos Ladner (born 1796), he fought with Andrew Jackson in the Battle of New Orleans in 1814, and he settled right here where we live. He got out of the war and settled in. We're east of the Wolf River. His brother John Ladner settled west of Wolf River. The people back in those days were ten miles apart, and then it was too close for some of them, and they'd go further off."

The second time I spoke to Billy Frank, it was not a recorded interview, so I can't give direct quotes. I was curious about the name *Keene* that Arie indicated was associated with the stock Billy Frank once owned. He told me about going hog hunting with his friend, Fred Keene, in Pitts, Georgia. Farmers would put out corn to attract the pigs and then catch them. He estimated the time period to be the early 1990s because he recalled driving a GMC pickup that was new in 1993. Billy Frank went back later to pick up a blue boar that Fred had raised in a pen. At that time there were lots of blue hogs both in Mississippi and Georgia, he asserted. He bred the blue boar with several sows from his Mississippi Guinea stock and kept a black boar named Andrew that lived to be almost twenty years old. He claimed that the blue hogs were blue from birth. They had a short body and short legs. They were real gentle, he said.

My third conversation with Mr. Brown was recorded in 2017. Each time we spoke, Billy Frank recalled slightly different memories. Billy Frank remembered that in the 1940s, his grandparents raised Guinea Hogs. His oldest memory was when he was five or six years old. "It was just open range and we would ride up on the horse and call them [the Guinea Hogs], and they would just come up like cattle. It was an open range, you know. They lived in the woods behind Grandma's house. They was raised just like the cattle and the sheep--- in the woods in an open range. Other people had them, too, you see. They had them earmarked. And everybody had their own earmark and when you went to mark the pig, so-and-so's pigs were marked. We did it with a knife. We didn't have notchers back then. With a pocket knife. It was my daddy's mark. That was in the left ear, and then there would be another one in the right."

I asked Mr. Brown to tell me more about the color of his blue boar. "He was dark blue. I would just call it blue. Gray, blue, same color, you know. They were born blue. If they were going to be blue, they were blue when they were born. There are still blue ones around. I think there are still blue hogs in Georgia."

I asked him again about Fred Keene. This time, he recalled that the Keenes lived in Abbevile, near Pitts. "Fred Keene, and his wife Suzy, we went away one time and brought some back. If Fred is still living, I bet he still has hogs. He had a big pen to keep them in, kind of like a hog lot. Two or three acres with a pond in it. Plenty of water. Now this has been 20 some years ago last time I was in touch with him."

After four years of searching for Fred Keene and reaching only dead ends, I decided to give it one more go in November 2018. This time I searched for Suzanne Keene on Facebook and reached out to one or more women with that name. Two months later, in January 2019, I received a response to one message. I had reached the correct Suzy! She told me that Fred was now seventy-eight years old and would be happy to speak with me. She included a phone number.

When I called Fred, I explained about my book and that Billy Frank had told me about their hunting trip in the early 1990s. Fred told me he had hunted in the area of Pittsboro, Georgia all his life. "In fact," he said proudly, "I am seventy-eight, but I am going hunting tomorrow for wild boar. I don't hunt anything but wild boar. I asked him to explain to me how he did that.

"Just turn the hogs out, and we go in on a four-wheeler and we turn a dog loose. And me and my hunting buddy throw the hog and put cuffs on him. That is how we catch them." I wondered how they restrained the hog. "Just catch them by the back leg first, and the dogs grab them by the ears and catch them by the back leg. The other person catches him by the opposite front leg and lay them on the side. We leave a dog on him until we get them in cuffs. It don't take but one minute to put the cuffs on him." It sounded like he had that technique down pat.

Fred continued, "We used to tie them down with parachute cord, but we don't do that anymore. We use mixed-up cur dogs to help us. They have a little Bulldog in them and a little bird dog. They are just mixed breed dogs, not purebred."

I asked Fred about raising Guinea Hogs. "I never raised any hogs," he responded. "Most of the time we just catch them and turn them back loose. Sometimes we eat one. We barbeque them. They are good on the barbeque. We just skin and quarter them up, wrap them in tin foil, and put them on the cooker and cook them for about eight hours. We cut them up and put barbecue sauce on them. Miss Griffins. It is the best."

I asked him about Guinea Hogs in his area. "There are not any of them left around here," he replied. "They were mostly black or blue. They were smaller. Shorter legs. We brought Russian hogs in here, wild boars about twenty years ago. Now most of the hogs have some Russian in them. The pigs will be striped like a zebra. And the black hogs will have gold around the neck. I don't know anybody who raises Guinea Hogs now."

The genetics from the Cowpen Creek bred boar, Brown's Blue Boy, moved forward through his daughter, Maveric Esther. Esther was obtained by Donna Dorminey of Solomon's Wisdom Farm in Kentucky and was registered via the Genetic Recovery Project in November 2017. A few of Esther's offspring were sold and registered before the ten-year-old sow's death in February 2019. Breeders with hogs that include Brown's Blue Boy's genetics include SWF MS Willie Mae belonging to Sandee House, SWF MS Lenah with the Robersons, Maveric PeeWee, Sprague Maveric Buster, SWF MS Breandon at Solomon Wisdom's Farm, BRM MS Memphis Minnie with Jessica Creighton, BRP MS Artesia's Pride with Sam King, and BRP MS Willie Dixon with Nathan Bundrick in Georgia.

Dan and Shirley Hale

Shirley and Dan Hale live in Plainwell, Michigan. They sold their last Guinea Hogs twenty-three years before I contacted them in 2017. They were inspired by the work of the AMBC and sought out minor breeds to raise. Through their research on minor breeds, now known as heritage livestock breeds, the Hales discovered those that the Guinea Hogs were a good fit for their farm. They raised a variety of heritage breeds, which were difficult to find in 1986 before the conveniences of smart phones, personal computers, social media, and internet search engines. Because of some technical difficulties with their phone line, I was unable to record this interview. However, I took detailed notes that I summarized immediately following the interview.

Initially, the Hales were most interested in Dexter cattle, another heritage breed. However, their youngest daughter, Nicole, was determined to add pigs to their homestead. The Hales came across a magazine article about Guinea Hogs and started their quest to find breeding stock. They researched the breed at a time when very little information was available, but their sources stated that the breed was a southeastern hog by tradition. The Hales consulted researchers at Michigan State University in Lansing for information on the Guinea Hog, described as a solid black pig. The Hales asserted that all of their hogs bred true to that solid black form, and had never heard of blue Guineas or of hogs with white spots or markings. They did, on occasion, have hogs born reddish that faded to black in a few months. They considered this variety to be solid black.

Dan and Shirley started their herd with purchases "from a broker in Michigan" who supplied them with two gilts, Wilma and Betty, and a boar, Fred (This trio was named after characters in *The Flintstones* animated series on TV 1960–1966). They kept a closed herd based on two bloodlines from their sows, who had different phenotypes. Wilma was a stocky sow with a short nose, while Betty had more length and a longer nose to match. Butchered offspring had hanging weights of 100–110 pounds. The mature sows weighed only 120 pounds. Fred was kept lean, about 140 pounds. He was a dependable breeder with a calm temperament. The Hale hogs were likely what old-time breeders call the "little-boned" variety.

In spite of close linebreeding, the Hales did not experience any inbreeding depression during the eight years they kept the hogs. The sows bred twice a year, and the Hales sold meat to repeat customers every three months from staggered breedings. They kept close to 15–20 sows in their herd from their two maternal lines. Many of the sows were half siblings, with Fred being the sire.

Dan had a tragic traffic accident in 1994 that resulted in his disability. The accident coincided with a very cold winter in which piglets were crushed trying to keep warm under their dams. That winter temperatures in Plainwell reached minus twenty-two degrees Fahrenheit. Dan's injuries, a brutal winter, and devastating losses of piglets contributed to the Hales' decision to disperse their Guinea Hog herd. Because they sold pets and meat more than breeders, Hale Mork (listed in the AGHA registry as Hesters Mork), purchased one year earlier by Annette Hesters, is the only hog from their herd to have known descendants. The Hales were thrilled when I told them that their conservation work did not come to a dead end at that point, as they had previously assumed.

When the Hales started their Guinea Hog project, Vietnamese Pot-bellied Pigs were all the rage. Guinea Hogs were crossed with Pot-bellies and also kept as pets due to their small size and pattern of slow growth. The Hales noted that overfeeding Guinea Hogs can cause obesity that negatively impacts breeding and produces too much fat in the meat. They used a custom mixed horse feed that was predominantly oats and corn, along with hay as winter forage. The hogs loved to graze, they reported.

The Hale legacy continues through Becky Mahoney and the conservators who purchase her carefully selected, smaller line of hogs. Names on pedigrees that indicate Hale (Hesters) Mork's lineage include Hesters Polly, Hesters Patti, Hesters Cracker, Hesters Petunia, Hesters Peter, and Hesters Carlos. Carlos, Polly, and Petunia are each both offspring and great-offspring of Hale (Hesters) Mork, and therefore more likely to represent his phenotype.

Annette Hesters

Annette Hesters was born Annette LeBlanc in 1941. The LeBlanc family settled in the area of Walkerton, Indiana in 1832. Annette and her husband married in 1959 and made their home in Walkerton. To this writing, they continue to live on the premises of Hesters Farm Log Homes, an educational venue with log homes and displays of heritage livestock for agritourism. One son lives adjacent to them and keeps Jersey milk cows, horses, and sheep. Their website states, "We pride ourselves in providing a unique country setting to help you get away from it all. We are centrally located, yet completely isolated from the hustle and bustle of your day to day activities." The Hesters family raised ten children there. Annette describes four of them as "homemade," and the other six were adopted.

Even at the age of seventy-seven, Annette maintains, with family assistance, a few Guinea Hogs and Shetland Sheepdogs. Her facility plays host to regular seasonal "rendezvous" events that return seasonally year after year. They also teach or host sustainable living workshops on cast iron cooking, blacksmithing, quilting, canning, Dutch oven cooking, and more. Annette's memory from breeding pigs as long ago as 1987 has faded, but she kept some breeding records and organized them in late 2005 when Arie McFarlen purchased hogs from her. I have a copy of those notes, obtained from Annette by Becky Mahoney (featured in chapter 11).

Annette told me that she discovered the hogs from an article she had read. It could have been the 1987 *Chicago Tribune* article that included an interview with Marcia and Paul Read, because Annette bought her first duo of Guinea Hogs in either 1987 or 1988 from the Reads. They owned Old Orchard Farm in Claysville, Pennsylvania. Marcia shipped the breeding duo of Iggy (boar) and Ziggy (gilt) to Annette. For three years, she bred them and retained two sows—Hesters Molly, and Hesters Iggy II. In 1990 she added a third sow obtained from Donna Watkins in Lexington, Illinois. Annette's notes say that she obtained a "Virginia gilt via Donna Watkins." I interpret that to mean it was not from Watkins' farm, but stock Donna imported from a breeder in Virginia. We do know that J. Frank Baylis had an active herd from 1982 to1994, so the gilt, named Ginger, may have been offspring from his stock.

In 1993, Annette changed boars. She had one delivered from Dan and Shirley Hale, minor herd breeders in Plainwell, Michigan. This boar, Mork, is recorded on pedigrees as Hesters Mork. Mork was bred to the Virginia hog Ginger to produce Hesters Polly and Hesters Patti. He was also bred to Hesters Molly to produce Hesters Cracker. Annette's notes indicate that there was a "spot" on Cracker. Mork was linebred to his daughter Polly to produce Hesters Petunia and Hesters Peter. Mork was also linebred to his daughter Cracker to produce the boar Hesters Carlos. For twenty-three years, Annette kept a closed herd. She told me she bred fathers to daughters and castrated the sons.

In 2005, Arie contacted Annette and offered to trade some stock with her. She had just started with the Guinea Hogs by obtaining a herd of breeders plus four litters from Stephen and Hollie Brothers in South Dakota. Arie traded two related gilts and an unrelated boar for Hesters Petunia, Hesters Cracker, and Hesters Carlos. The piglets Arie obtained from the Brothers and traded to Annette included only one litter that was unrelated to the other three. That litter was born July 26, 2005 from dam DNC Esmerelda and sire DNC George. The litter included two boars, Brothers Brewster NR025 and Brothers Bronson NR28.

One of those brothers was probably Annette's new boar from Arie. The gilts traded to Annette were sired by SCZ Bullwinkle NR002 from Celesky lineage at Sedgwick County Zoo. The gilts' dams were either Brothers Penny NR005, Brothers Chunky 2 NR013, or Brothers Patti (Pearl) NR004. Annette retained an old sow from her Iggy/Ziggy Read line (descended from Molly or Iggy II) to include with her new hogs and employed linebreeding from that point forward until 2015. Annette selected for smaller sized hogs, about 120 pounds at three years. This is consistent with the size of her boar named Mork.

Don Oberdorfer once told me what he recalled from 2005. "What do I remember about the Hesters herd? When we were trying to do the assessments of what Guinea Hogs were still around, Annette Hesters was one of the breeders that still had them. She said she could help us out by selling some if we needed them. She'd had them a long time and wasn't willing to give them all up. That was before the organization was formalized. Arie had the desire and means to go get them."

By 2015, Annette was ready to downsize her herd so she could go south each winter. She contacted The Livestock Conservancy for advice on placement. Staff there contacted me, and I reached out to Becky Mahoney to engage with Annette. The hogs Annette sold to Becky in 2015 became part of the Genetic Recovery project. Becky has been managing and selecting the best hogs from the herd.

Donna Watkins

The name Watkins as an early Guinea Hog breeder was featured in an article written by Miranda Bryan for The Livestock Conservancy, titled "American Guinea Hogs — Pedigree Analysis and Breeding Recommendations." In a chart listing the foundation boars for the breed, "Watkins Stock" was listed. Over the years, as I searched for the foundation bloodlines, I kept my radar up looking for a confirmation of that name in the foundation herd.

When Becky Mahoney took over Annette's herd, she worked with me to prompt Annette's memory in regard to records on the hogs. Eventually, Annette located her old records. One note referred to a gilt she purchased in 1990. "Virginia gilt via Donna Watkins, Lexington, Illinois. There was the Watkins connection! The wording of the note makes me think that Annette contacted Donna for hogs. I suspect that Donna had a gilt from Virginia that she was willing to sell, and was the seller but not the breeder of the hog. J. Frank Baylis of Bayshore Kennel and Barn lived in Virginia, started breeding in 1982, and was still active until 1994. The Virginia gilt named Ginger could have come from Bayshore Kennel and Barn, from J. Frank. He had a website and a brisk business in breeding stock. He told me that he sold his stock far and wide.

Going back to interviews with Kevin Fall and Don Oberdorfer, I found references from each of them regarding how Mark Celesky obtained his stock. Both mentioned "a lady in Illinois." When I spoke to Mark about his brief foray into Guinea hogs, I asked him if he ever got a pig from Donna Watkins. His response was "that sounds familiar." For those who wonder about fuzzy details like this, it is important to realize that, before the association and formal registry was formed, these things were not being tracked. When someone made a trek to get the last few Guinea Hogs, they often took turns picking up hogs, delivering them, and trading stock. They often didn't travel to the state themselves and meet the breeder in person. Kevin Fall alone made countless trips, from the way breeders tell it, traveling across the country purchasing entire litters and dispersing them to conservators. These people were caught up in the moment and did not think that these details would ever matter to anyone in the future. There is a possibility that the Celesky hogs have a trace of "Watkins" bloodline in addition to Setty, so it is worth a mention here.

I was unable to contact Donna Watkins in Lexington, Illinois. My belief is that she has passed away.

Marcia Read

The Read bloodlines, sometimes-misspelled *Reed*, were referenced in documentation about the Guinea Hogs, but locating information on Marcia Read was elusive. I became motivated to find her after interviewing Annette Hesters and discovering that Annette's first duo of Guinea Hogs, Iggy and Ziggy, were bred by "Marsha Reed" in Claysville, Pennsylvania. In an interview with Callene Rapp at the Sedgwick County Zoo, she confirmed that Marcia Read of Old Orchard Farm in Pennsylvania sold the zoo a Guinea Hog in the 1980s. Gabriella Nanci told me that she believed one of Marcia's daughters had Dexter stock listed on the Dexter cattle website. I went online to chase these leads and started finding information about the Reads' activities.

Old Orchard at Stockdale Farm in Claysville was situated on Highway 321. The *Equine Now* website listed the farm as a breeder of Haflinger horses, Dexter cattle, Jacob and Tunis sheep, miniature donkeys, and Pygmy goats. I found a *Chicago Tribune* article that placed Paul and Marcia Read at the tenth annual meeting of the American Minor Breed Conservancy near Chicago in October 1987. The couple had been quoted in the article. I think, in the end, I found the elder Read daughter through the Dexter website and contacted her first. Then she introduced me to her younger sister, Jessica Magby. Jessica had been more involved with the Guinea Hogs that her sister.

I interviewed Jessica in 2017. Jessica was "just a kid" when her parents kept the Guinea Hogs, but she remembered them fondly. Jessica told me why she wanted to speak to me before I called her mom. "Some of her information might not be good." Jessica said. "She might give you answers that aren't true. She had a head injury, and it's permanent. I thought maybe you could talk to my dad first, and then if you want to talk to my mom and kind of filter through her information. She loves to talk to people."

Jessica told me Marcia was born in 1949. At the time of our conversation, she was sixty-eight. Jessica recalled that their original stock of Guinea Hogs came from Idlewild Park, near the Pittsburgh Zoo. Apparently they had a petting zoo, but that has since been converted to a water park. Marcia developed a friendship with the staff there. She loaned them Pygmy goats and Dexter calves for summer displays. When the park was dispersing their hogs, they contacted Paul and Marcia. That was how Old Orchard got started with "about four" Guinea Hogs. Marcia was interested in homesteading and in easy-care livestock. She was active in the AMBC. Jessica stated that the original hogs from Idlewild Park were solid black, but one or more replacement hogs from the Southeast had an insignificant spot of pink on their noses. They did not have white on their feet. Jessica told me the minor role the hogs played at Old Orchard.

"You may have caught on that at Old Orchard, we did a lot of stuff," she began. "The hogs are probably what we invested the least in. We never had very many. My mom identified as the sheep person, and we were very big in Pygmy goats. Letting go of the hogs was not really much of a choice. My parents split up, and that created some financial distress. We only had one breeding pair at the time. I'm the one who talked Mom out of getting rid of them because they were not breeding.

"I think the boar must have been sterile, because I remember him being kind of flat in the back. I was old enough to remember them, but it was twenty years ago when I was in college (late-1990s). I was primary caretaker of the pigs because my sister was scared of them. I was pretty young, but as kids, we did most of the animal care. I don't remember having two boars at one time. And I don't think we ever had more than four sows. There were normally two. We probably got them around 1980, when I was five or six, so we would have had them about nineteen years.

"At one point, they almost lost the entire herd to a barn fire. They only saved one piglet, and then they went and got more breeding stock from somebody in Georgia or Alabama. They had to go pretty Deep South, and they found a guy that had to be raising them for generations."

When I spoke to Paul, he narrated a slightly different story than his daughter's. He said the fire happened after new stock was obtained. He told me they were raising only the stock they had until they located a man in Georgia. He thinks it was in the late '70s or early '80s. "I remember Marcia and I driving a truck from Pittsburgh down to Georgia, and finding, in the back woods, at an old beat up trailer, a guy that had raised them for generations. And going down and buying a new boar, and a new sow, I believe, and driving all the way back to Pittsburgh. I don't even know how we found that guy down in your neck of the woods. And it was not a picture postcard of Georgia. There was junk everywhere around. Some old guy came out in a white sleeveless undershirt, and just a rough old boy, but he knew these hogs. Everybody used to raise them. He said everybody had them in his area. I'm surprised that there's not a pocket somewhere in Georgia that you can still find them. I guess they've gone the way of the family farm."

Gabriella Nanci also remembers the fire, and how only a beloved gilt named Francine survived it. Gabriella said that Marcia bought back a duo of pigs she had sold, even though the boar was a half-sibling to Francine. She does not know if this event occurred before or after the Georgia stock was obtained.

Paul continued, "I remember that the breed was very docile. The boars were very docile. They had a purpose being lard hogs, and they were resistant to hog cholera. At one point, someone from the government, maybe from the Department of Agriculture, contacted Marcia to see if we could provide so many pairs of Guinea Hogs to be sent to some island like American Samoa or someplace, to help repopulate the hogs and confer tolerance to hog cholera."

The common lore was that in a time other hogs were spreading disease, these homestead hogs stayed healthy, even in the hot and humid South. The transfer was never made, because the Reads did not have the numbers the USDA needed to send hogs to Samoa.

Jessica recalled that her parents donated pigs to a school called Green Chimney in New York State. It is still in operation in Brewster and with a satellite location in Carmel, both non-profit schools with therapeutic programs for special education students. The original school was founded in 1947 and specializes in animal-assisted therapy. They keep over 200 animals that partner with residents for education and therapy. Jessica remembered, "We got one of the pigs back and she was a beast - she was so large she could barely walk. And of course her name was Miss Piggy. They loved her a bit too much, and we had to get the weight off her. I don't know if I'm remembering correctly, but I think Paul Newman met one of our pigs there. I heard everything secondhand."

In 2018, I contacted Maureen Doherty, farm manager at Green Chimney School. Maureen was working there in 1996 and remembers an African Guinea Hog named Stacy. "She was super sweet," Maureen said. "The kids really enjoyed spending time with her. I do remember employees talking about her being donated. Here at Green Chimney, especially at that time, the kids would come and care for the animals. The pigs in particular, they took real extra care to make sure they were comfortable, and they would pet them and make sure they had nice mud to cool off in."

She proceeded to tell me a story about Stacy. "We were out in what we called the center of the farm. We had her out there for the summer. One day we came out and she was gone. The gate was open, and she was gone! 'Where is she? Where is she?' We looked everywhere. Everywhere! We couldn't find her, and we were panicked! To the point where someone said, 'Somebody stole her!' Well, Stacy had jimmied the gate, or someone had left it open and she got out. Not really sure how it happened. But she made her way over to the manure ramp, up the ramp, and the dumpster was just full enough that she walked down the pile of manure. And there she was, in the dumpster, at the bottom, completely sacked out, like she had just had the best time of her life all night. So that's my hugest memory. And the kids, they just laughed so hard. We had to all work very gingerly to get her back up that manure ramp, and then she walked down. That's a funny story about Stacy."

Maureen also recalled that the children loved to scratch under Stacy's belly and that she would lay down for more attention. "They would scratch where the skin is so soft, behind the ears." Stacy was the only African Guinea Hog at the school, but they also kept GOS, Tamworth, and Vietnamese Pot-bellied Pigs. I was delighted to read an article about Green Chimneys in the March 1, 2019 *New York Times*. It focused on how the children's lives have been changed through those animal interactions. The featured photograph is of two boys greeting a friendly hog. Of course, I thought of Stacy.

I got the impression from talking with Jessica that her mother had been very independent in her farming years. She agreed. "She could be very passionate about her rare breeds. I honestly don't know which was her favorite. I would say that, as a family, we had the greatest presence in the breeding of Pygmy goats. Mom had an amazing vision. Every breed that she got became very popular. We had the first Haflinger stallion in the state, and they are ubiquitous now. The Haflinger horses came from New England, I believe. The person who had them was actually logging with his Haflinger horses. They had been recently imported. The stallion came from the same place that has the Lipizzaner stallions that we know today.

"My dad can definitely tell you where the Dexters are from. I think they bought out a herd that was going to be sold off. Dexters are now in mass numbers, and miniature donkeys were rare and exotic and then exploded. I know the Conservancy still watches them, but there are tons of them, and you can get them for free. She definitely had a vision, but she was not always the best at following through," Jessica stated.

"We had one of the first big Pygmy goat show herds on the east coast," she continued. "They were dominant, and my sister became a judge. To this day, I'm not sure why Mom did what she did. We lasted so long with the Dexters. She is very much a purist. She cared about them. And they make money for you. They're kind of like the pigs. You have to have them, and they do their job. Little black animals that work very hard and never get sick. I am not a cow person, but the Dexters mattered a lot and we raised them for so long that I never remember not having them. So yeah, she strongly identified with her cows. I know it was horses that initially drew her to farming. It was a bit of a hippy *return-to-land* thing. And that was abandoned. She loved the allure of the show ring. That is probably one of the reasons the pigs weren't so important. You didn't show them.

"Our first Jacob sheep is tied to the Rockefeller family somehow....Our Tunis sheep, they came from a Pennsylvania breeder. We had them for years, and now they are popular. I shear sheep part time. Those are one of the first flocks that I did. I shear small flocks. I drove over an hour one time and pulled up and saw some little hairy pigs. I said, 'Oh, you have Guinea Hogs?' And they said, 'You know Guinea Hogs?'"

When I spoke with Paul, he lamented that over the years he has seen the heritage breeds change from what they were in the 1980s. They were either larger, smaller, different colors, or polled. He was concerned with choices breeders had made that impacted the breeds. In the case of the Guinea Hog, he had seen people breeding them smaller and smaller in response to the pet pig fad of the day, and crossing with the Vietnamese Pot-bellied Pigs until the pure Guinea Hogs became extremely rare. Paul and Marcia had both been involved in various minor breeds boards of directors. He felt that egos got in peoples' way, as everyone believed that their strain was the best.

Paul was openly upset about what he has seen in regard to genetic changes in the breeds he had raised. Haflinger horses are no longer relatively small, he claimed. Angus cows are "twice the size" as they were, Southdown sheep, always small, have been miniaturized in the Baby doll type. He is concerned that the same changes may be taking place in the Guinea Hog. White markings were considered a sign of inbreeding, he reported to me.

Paul remembered a story about the docility of the Guinea Hogs. "I recall that we had a sow that was delivering, and we didn't know that she was delivering. And the boar was in with her. And our good friend who was a vet tech was really impressed because the boar didn't eat the offspring. Instead, he was very carefully walking around these little babies that were coming out, that were small enough to fit in the palm of your hand."

I had planned to interview Marcia. However, a few weeks after my interview with Jessica, I read her Facebook post announcing that her mother had passed away. There was a beautiful photograph of young Marcia, maybe in her late thirties or early forties, in the kitchen with a teenaged Jessica. They shared smiling eyes and wide grins. Jessica wrote a caption on the photo. "Some things I learned from my mother. 1. You should always buy the book. And the horse. And maybe the goat. 2. Farming is awesome. 3. Few things cannot be improved by Def Leppard and coffee." I'm sorry I never got to speak with Marcia. I could tell we might have been great friends in the right circumstances, and had much in common regarding a passion for heritage breeds.

The Sumrall Family

Members of the Sumrall family who raised Guinea Hogs in Mississippi include Brent Sumrall (1979-), Gary Sumrall (1954-2016), John William Sumrall (1921-1986), and Harmon David Sumrall (approx.

1895-1960). Brent, Gary, and John all experienced the hogs as young children. The Sumrall families in succession raised Guinea Hogs continually prior to Hurricane Katrina, a storm that devastated the area in 2005.

I first heard about established herds in Mississippi from Don Oberdorfer and Kevin Fall in 2013 and 2014, and I had seen the "Sumrall Boar" listed in Miranda Bryan's chart published by The Livestock Conservancy in 2014. I didn't learn about the Mississippi-Sumrall connection until I got a lead from AGHA member Rico Silvera in December, 2014. He knew I was doing research on a book about the history of the hogs and was concerned about regionalized genetic pools in the breed. In his search for diversity, he contacted Arie McFarlen.

Arie returned Rico's call and told him she had distributed her stock in 2014. She had kept just a few of the older boars and wasn't breeding anymore. She gave him the name and contact information for Gary and Brent Sumrall in Mississippi. The Sumrall family was breeding again, but Brent told Rico all they had for sale was little-boned stock purchased locally from a family with registered Guinea Hogs. Gary actually had other stock, I later found out, but was not ready to sell any offspring because he was working on recovering a specific type from old family lines. He was keeping everything to grow out before selecting the keepers. That is why he had only the little-boned stock for sale. Rico could not find out who had taken over Arie's herd or where she distributed them. He gave me a contact number for Gary's son, Brent, also known as Buck.

I got off the phone and sent an email to Brent. To my surprise, he responded to me within a few hours. He told me a bit of his family history. "My great grandfather, grandfather, and now my father raised the Guinea Hogs for years. We let Arie McFarlen have the ones we had left when we were down to a few. Arie promised my father anytime that we wanted to get our line of Guinea Hogs back that we could. And back last year she called and told my father she had our old bloodline of hogs that she had preserved. We went and picked them up, and now we have 50 of them. We haven't sold any of our own bloodline we got back from Arie because we are trying to build our original bloodline herd back.

"My father had looked in the Market Bulletin for the past 30 or 40 years to find replacement hogs just to get some new bloodlines, and never saw any. The last time we brought in hogs was in 1961. But around 2011 he found someone advertising the little-boned Guinea. He went to get them and brought two boars and five sows. Then, when Arie called us to get our old stock we got them. Our old stock is the big-boned Guinea crossed with little-boned Guinea my grandfather got in the 1960s. The big-boned Guineas, I think, is pretty much non-existent now. But ours for the last 40 years have been a cross of big-boned and little-boned. My father knows all about them. I was raised all my life with them. The hogs we have top out at 300 pounds. The original big-boned Guineas would weigh, I would say 400 to 500 pound size. But it's impossible to find any of the original big-boned Guineas.

Buck continued, "Yes, my father would be happy to talk to you. I know Arie told us that she didn't sell any of our old stock and bloodline. She strictly used them on helping to bring the bloodline back. Up north, the only other people that had our bloodline was a zoo up there. I can't remember the state, exactly, but they are our original bloodline that we have except for the little-boned Guineas my father purchased around 2011 [from the Shirley family]. As a matter of fact, our [1961] little-boned hogs originally came from Georgia. The old Georgia stock. My grandfather had some of the blue Guineas, but we don't have any of them. The blue Guineas was just like the black Guineas as far as having a big-boned and little-boned type. My father will be happy to talk to you. We have had the hogs for generations and will tell you as much information about them that we know."

This was a very exciting connection! I reached Gary by phone on January 6, 2015. Gary explained that he was never interested in joining the AGHA. His family had raised the breed for a hundred years before the organization was formed. He said, "I want you to see my hogs. I've got a big-boned Guinea sow that is supposed to have pigs any time now. And then I have two little-boned sows that is bred to the big-boned Guinea male. The picture that I sent to you, that male [Maveric Hansel]. And I am anxious to see what they produce. I don't know how you are on your hogs, but I will probably have a few to sell before too long on the half big-boned and little-boned. (Gary was linecrossing his family Sumrall line from Maveric with stock he got from the Shirley family to avoid excessive inbreeding).

"My daddy was inbreeding his hogs," Gary told me, "so he had to have another male. And he sent his brother and my oldest brother and my uncle's son to Griffin, Georgia in 1961. I can't think of the man's name. He was ninety-something years old. He had the blues and the blacks. They went over there and got some of them. Ma'am, I'm proud that you called me, but I have to go right now."

Gary explained to me that his family did not take pictures of their old stock. It was just so common for his family and neighbors to keep hogs, and film was too precious to use on something so mundane.

"Keeping hogs was their livelihood, you know." Gary said. "But you can look at ours and you see what you get, you know what I mean?" Gary was proud of his hogs, and for good reason. They were fine specimens of the breed. "That's just it in a nutshell," he concluded.

I was excited to take a road trip to Mississippi to visit with Gary Sumrall in person and finally meet his hogs. The Sumrall family's mailing address is Laurel, but the community they live in is called Sharon. My live interview took place on November 18, 2015. Gary told me about the three generations of Guinea Hog breeders in his family, starting with H. D. Sumrall, his granddaddy. "H. D." stood for Harmon David. Gary did not have a birthdate for him, but Gary's father was born in 1921, so a good guess would be that H. D. arrived around the turn of the twentieth century in 1900. H. D. was the Jones County surveyor for years. He had a farm in the Soso community about fifteen miles from Sharon. "But he had cows and they done a lot of crops with the mules. And he had, what happened, he turned it over to my daddy after my grandmother died. And then Grandpa remarried, and he moved to Soso. He lived in town."

From Gary's earliest memory (he was born in 1954) the family kept Guinea Hogs. "Daddy worked on road construction for the county," he stated. "He had cows and mules and horses and Guinea Hogs. They'd raise corn, peas, and butter beans." Gary remembers taking a trip out of town with his daddy, John William Sumrall, to get a new boar for the herd. The year was 1961, and they were heading to Griffin, Georgia. "We got some little-boned Guineas. Daddy had the big-boned Guineas, and he got a male and two females." He showed me a faded Polaroid picture of a trio of young hogs about three or four months old. "This guy that had 'em, he was probably in his early nineties, and he was selling out. It was the last of the herd. And Daddy said he had close to one hundred head. He had a pretty big herd of them. And this is some pictures of 'em, of the Georgia stock hogs. They're real thin-haired hogs. Daddy's was thick haired. But Daddy crossed 'em up with his big boned. They were all crossed. Over the years, he crossed 'em up."

Gary continued about his daddy's big-boned hogs. "Real thick-haired, and they were big, pretty big hogs. They'd weigh, the males would weigh 500 pounds by three years. For butchering, Daddy always got his up to about 250, 300 pounds. And every Thanksgiving, that's when we would butcher. They would be about one year old or a little older. They grew quick, they did. You don't find them like you used to. They're a smaller type now. Most of the ones I've seen have been a smaller type. But see, you got to a point that you couldn't find 'em. And that's the reason Daddy went to Griffin, Georgia and got him a male to cross to the last he had. At that time it was hard to keep from inbreeding, you know. But at that time, he just couldn't find the old type anymore. They got harder and harder to find, you know what I mean. We had to breed what we had.

"And Daddy, all he would fool with was Guinea Hogs. They were real easy to handle. My old male out here, he's just as easy going as he can be. Daddy would usually feed shell corn to his. He had 'em on the outside. He'd throw a coffee can, two-pound, three-pound can, and throw it to 'em. And they grazed just like a cow, uh-huh. Daddy usually kept around six sows and usually a couple of males, plus the pigs. They mainly grazed like a cow. And when Daddy died, I promised him I'd take good care of his hogs." Gary was teary eyed when he told me that, and I'll admit that I was, too.

We talked a bit about the devastation in their area after Hurricane Katrina hit in 2005. Gary lost his old boar, and had only three older sows left. He couldn't locate a replacement boar to breed them. Arie was in Louisiana for a presentation. "And I got in touch with her, and I let her take those three sows. And the deal was, she would let me have some back when I was ready for 'em. We had rental property and Katrina done us in pretty bad. We didn't have time to fool with 'em, and that was the deal I made with her.

"We got on our feet after a while and found some people in Brookhaven, Mississippi. Their old hogs were registered, but we never registered ours. And then, when Arie got ready to get out of it, she called me, and I went up there. She had moved to Huntsville. Well, she called and said, 'Come up here and get you some hogs.' That was in February 2013 that we went up there. She had, I'm gonna say sixty or seventy head. And she had them all sold but the ones I got. She kept mine separate....But now, she didn't register her hogs, either. I don't know who she let have the rest of her hogs. But somebody came and bought all of them."

Gary continued, "We bred those little-boned hogs from Brookhaven to the ones from Daddy's line that I got back. I got one little-boned Guinea sow. She's got a long nose. My daddy had some long-nosed Guinea Hogs. They'll raise pigs like a rabbit. I mean they'll have pigs.

"Now, Daddy had got some blue Guineas from the guy in Griffin, Georgia. But he didn't keep 'em that long. They just wasn't as active as a black Guinea, and he didn't care for 'em that much. I know Arie got a blue Guinea male from somebody down in south Mississippi. And she crossed him up with a black sow and she was hoping to get the blue started again. But they're pretty extinct, the blue. I don't know of any. They're not as good as the black hogs to me."

Arie had previously told me about the three old sows she got from Gary after Katrina. Two of the sows never farrowed again. The third sow, Sumrall Bobbie Sue, had a single litter, out of which Arie saved a boar she named Maveric Charles Sm3. From this one boar, she made several matings that imparted 50 percent of the old Sumrall line to his offspring. She saved these genetics over the next several years, then returned them to Gary as promised.

Gary gave me some words of advice. "On your sows, watch your weight, because that determines the number of pigs they have. Because if they're overweight, and you probably already know this, it'll cut your litter down a lot. And you need to average eight pigs at least to make it worthwhile fooling with 'em. They get fat too quick. You gotta watch."

Just two weeks after this interview, Gary was admitted to intensive care. His prognosis was not good. About a month after that, he succumbed to cancer, leaving the family who loved him bereft and with nobody able to care for his beloved hogs. Several conservators would step in to keep the genetics moving forward, both to honor the Sumrall family legacy and to benefit the overall registered herd in the United States. If you see a Guinea Hog pedigree with a hog named (farm prefix) MS (name), this is a descendent of Sumrall Bobbie Sue, and includes genetics from the Sumrall family legacy.

Part VII
Expanding the National Herd

Chapter Ten
Genetic Recovery Begins

> "Breeds serve as the main reservoirs of the genetic diversity within species. Half of the biodiversity of most domesticated species is shared across breeds, while the other half is unshared and is instead contained only within single breeds. The consequence of this is that losing breeds means losing genetic diversity, because by losing a breed the species loses the genetic information that is unique to that breed."
> *Dr. Phillip Sponenberg, Alison Martin,*
> *and Jeannette Beranger, Managing Breeds*

My Story and my Process

I did not begin researching and writing a book because I was an expert on this topic. It was quite the opposite, in fact. I started farming at age fifty-seven, after retiring from a thirty-three-year career in elementary education focused on struggling learners. I had no experience farming, but I knew how to research, make observations, and ask questions, and I believed in lifelong learning.

I had been raising egg-laying poultry, a milk goat, three breeds of heritage meat rabbits, and a small flock of Gulf Coast Native sheep for three years when I heard about the Guinea Hogs in 2013. I have always loved pigs, but did not want a 900-pound boar or an angry 600-pound sow to deal with. I had only eight acres of fenced pasture land, and the sheep used that for rotational grazing. The Guinea Hogs seemed to be a perfect match for a small landholder and beginner such as me.

It found it surprisingly difficult to find information on the breed. The AGHA website had a bit of information, and so did The Livestock Conservancy. Only a few breeders of registered Guinea Hogs had informative websites at the time and each one I found repeated the basic information found on the AGHA and The Livestock Conservancy's websites. The Maveric Heritage Ranch website seemed to have the most information, with details about sources of herds, names of lines, and characteristics of strains. It was just enough to pique my curiosity, and I was intrigued.

I am not the kind of person to easily quit or become discouraged when I hit a bump in the road. Rather than complain and mope, I tend to look for solutions and actions I can take. I had ascertained that nobody had written the book I needed to read. Therefore, I needed to write the book so I could read it! I completed my doctoral dissertation in reading education in 2004. I was well prepared after that challenge to seek out sources, record their stories, ask questions, transcribe recordings, and synthesize information in narrative form. I began to seek out experienced breeders online and announced my intention to write a book. I knew next to nothing, and was ready to soak up anything and everything like a sponge. I had no idea what I would learn or where this process would lead.

My first concept of the book had the working title, *Hogs with Heart: Stories and Care of the Guinea Hog.* My idea was to gather stories about the hogs from old-timers, and interview or survey breeders with experience to find out how to feed and care for the breed. I also wanted to interview chefs to find out what made the meat so special, and include recipes for pastured pork dishes. I was clueless about rendering lard and making soap and other products about lard. That book I thought of as *Luscious Pork and More: Using the Guinea Hog from Nose to Tail.* Then, I thought, after years into the Guinea Hog research, perhaps I'll manage to learn something about the genetics and family strains in the breed. That would surely be the last book, though, because I may never grasp that kind of knowledge. First I wanted and needed to read the book about how to care for my new Guinea Hogs.

I posted on the American Guinea Hog Yahoo Group that I was writing a book and looking for people to interview, especially those old enough to remember the pigs before there was a breed association. Every once in a while, someone would offer a gem. I started interviewing AGHA members here and there and soaking up what they shared with me. Then, one day, I was contacted by Lisa Naumann. Lisa had been breeding Guinea Hogs for several years when we connected in September, 2013. She very helpfully provided me with information that led me into the genetic lines and provided me with names of some key people. One thing led to another, and I was on my way.

An Email from Brent

In Chapter Nine, I introduced you to Gary and Brent "Buck" Sumrall in Mississippi, and told you about meeting Gary at his family farm. This is the email I received in October, 2015, that led to that visit.

"Hey, Mrs. Cathy. I hope your family is doing well. I wanted to tell you my dad is going to sell many of his hogs due to health reasons. We have about thirty of them. We have many meat hogs. Daddy is also selling the big boar and the sows." In a follow up email, Buck told me his father wanted me to have first pick of the herd and for me to coordinate any further placements for the hogs.

I was happy to help the family out to place the hogs carefully. After some initial plans fell through, I contacted some breeders in Mississippi. They agreed to take some of the hogs, even though they were unregistered. I was very interested in placing "Big Boar" and "Sadie" who were a mature breeding pair bred at Maveric Ranch. I had seen their photographs and did not want them going to a sale barn. I was not equipped to transport or keep a mature duo on my farm, but I did arrange to purchase a weaned boar shoat and a gilt.

One week after the call, I started a Facebook Group to discuss genetic lines in the Guinea Hogs. It has changed names once or twice and is currently "Registered American Guinea Hog Genetic Discussions." I also initiated a teleconference book study of the first edition of *Managing Breeds* by Sponenberg and Don Bixby. As most teachers can tell you, if you want to learn something challenging, the best thing to do is to teach it. I learn by discussing and summarizing and felt like the book study would be the best opportunity to surround myself with others interested in the topic. It was a nice group. We learned together, and there was a core of about four or five women who participated in the call.

I contacted my friend and neighbor, Pam Allgood, to see if she would like to share the drive to Mississippi with me. Pam is always good for an adventure, if she's not already on a trip to Europe or dog sledding in Alaska. She didn't even hesitate. We set a date to coincide with Gary's weaning of the boar shoat. One week before Thanksgiving, 2015 we were on our way.

Journey to Mississippi

Pam and I left Elberton, Georgia in my Prius V early in the morning on November 17, 2015. We had a seven-hour trip ahead of us. I rarely left the farm overnight, unless it was for a farming conference or to pick up livestock, so I planned a trip that would be fun for both of us. We put two of my largest animal crates in the back of the wagon. The Prius, which I referred to as my "truck," had previously transported a young ram, several ducks, poultry, hay, and more. On behalf of my non-farming friend, I was a bit nervous about returning with potentially stinky cargo for the trip home. I used my *Good Foods* app to find a

GENETIC RECOVERY BEGINS

farm-to-table restaurant as we crossed the Georgia-Alabama state line and the time zone from Eastern to Central time.

Garfrerick's Cafe in Oxford, Alabama was a pleasant surprise. The cafe's mission, "connecting people with their food," was similar to our approach in producing food on the farm. I, too, strived to connect my customers with their food. Chef David Garfrerick has his own farm where food for the restaurant is grown and picked, then transported a short distance for cooking. We had great service and a good, fresh meal for a reasonable price, in a casual atmosphere.

It was a warm, sunny day in Alabama. As we drove west on I-20, the weather became even warmer and more humid, in spite of the November date. We entered the state and stopped at the Welcome Center. It was the most attractive and well-decorated Welcome Center I had ever seen. The grounds were planted in magnolias—both the state tree and the state flower, although they were not blooming that late in the year. The main entrance to the center was a columned brick building with a tin roof, green shutters, a grand door and roomy front porch. The inside was decorated with antebellum antiques, and a pot of coffee was brewing. The floor gleamed, and there were brick archways in the ceiling. It looked like a cross between a museum and a very grand train station. We were transported to an earlier, slower time.

I had booked a room at Wisteria Bed and Breakfast, which is on the National Register of Historic Places in the heart of the historic district in Laurel. The price for a suite in this lovely lavender-painted home is less than most Hampton Inn stays, and includes a homemade breakfast served in a sunroom. The house was built in 1901 for a timber baron named Silas Wright Gardiner who was one of the founders of the town. The Schneider family had lovingly restored it and filled it with period antiques. The front doors have leaded glass panes, and the parlor's elaborate drapes are replicas of those hung in the blue room of the White House. We arrived at dusk, as the days were getting shorter. There was a warm breeze, and it was a surprisingly balmy eighty-six degrees. It felt like a July evening in Georgia.

After checking in, Pam and I went into town for a seafood dinner at the Blue Crab Grill. During the drive, and at lunch and dinner I had been filling Pam in on the history of the hogs and the missing genetics piece. I explained why this was an important journey for my book research and for my farming adventure. I was anxiously looking forward to meeting the Sumrall family and recording their history for future generations. I had no idea what to expect. I checked my pocket recorder to see if it was working properly.

During the night, it began to rain, and rain hard. Thunder, lightning, wind, and rain woke us up during the night. We had planned an early morning start but had to wait until nine o'clock for the weather to clear up. The unseasonably warm front must have brought this in. While we waited for things to calm down, we had a lovely meal with chicory coffee prepared and served by our gracious hosts. While eating our breakfast of fresh fruit, scrambled eggs, sausage, and bacon, we learned a bit of local history from our hosts. The rain eventually stopped, and we left for our main event. No more fun meals on the road.

There were flash flood warnings, and we drove by some flooded streets and overflowing gullies. We left the town and headed to a rural area with windshield wipers running and headlights on. Luckily, I had packed my muck boots.

When we arrived at the house, Janis and Brent Sumrall greeted us warmly and offered us something to eat, but we declined. I had fresh batteries in my pocket recorder and was ready to spend some time with Gary. Janis told me he was having a good day. Gary was smiling and propped up on one side on a couch in the family room. He was half reclined, not sitting up. We talked for about an hour. Gary answered all my questions and told me a bit about the hogs. His quotes are sprinkled throughout this book. The neighboring Mississippi breeders were going to pick up some young shoats soon, but I was unsuccessful in arranging a pick up of thirty feeder hogs. Plans fell through at the last minute. I was concerned that I could not find a home for Maveric Big Boar and Maveric Sadie, as Gary called them. But Gary assured me that someone was coming to get the rest of the herd.

Gary told me that the seven-week-old boar I was bringing home looked like his father's old type. He was very proud of the young boar. It would be the last one that Gary selected from a litter. He tried to explain how the hogs were related, but all of the hogs had names like "Big Momma," "Big Boar," and "Little-boned Boar." It was hard to follow and keep them straight, especially since Gary was too weak to walk me into the barn and point to the hogs as he described them.

After the interview, Brent took me out to the barn to pick up my stock. I was able to see the meat hogs and the little-boned boar that came from the Mississippi stock. Soon it was time for a long drive home. Pam and I returned with an eleven-month-old gilt that was on the petite size but had good conformation. If she had been any larger, she wouldn't have fit into my large dog kennel. The newly weaned boar rode in the smaller crate. We'd be unloading in the dark, but my farm intern and I had already prepared the quarantine area.

Right after Thanksgiving, Gary was put into intensive care and the hospital. He passed away early in the year. I was honored to have met the Sumrall family, record Gary's story, and help out in some small way.

Surprise from Indiana

Just eight days after I returned from Mississippi, I got an email from Jeannette Beranger of The Livestock Conservancy. She wrote that "a lady in Indiana who has been raising Guineas since the early 90's" had contacted her. "She started with stock from Marsha Reed [Marcia Read] in Pennsylvania and about ten years ago added some of Arie's pigs to the herd," Jeannette reported. She had two adult boars, three adult sows, three gilts over a year old, and four six-month-old shoats—two male, and two female. The breeder had told Jeannette that she was getting too old to manage the herd. "I think they could be important since they have been a fairly closed herd," Jeannette wrote. "I don't have any record of Marsha's stock in PA, so I don't know who they are related to." Light bulbs went off in my head. This was the *Indiana* herd I'd been told about!

Becky Mahoney, who happened to live in Indiana, was one of the regular members of my *Managing Breeds* book study. Jeannette gave me the go ahead to contact both Becky and the breeder, Annette Hesters, so I could collect a history of the herd. Becky agreed to stop by Annette's farm and assess the hogs. It turned out that Becky and Annette were separated by a mere ninety-minute drive. By December 2nd, Becky had already driven to Annette's place, chatted with her about history, and returned home with the whole herd. She was keeping one young boar for the winter only, to be returned to Annette in the springtime along with a gilt from Becky's herd so she could continue breeding when spring arrived.

Eventually, Becky collected mre information on the Hesters herd. The process involved several interviews with Annette, asking appropriate questions, and checking facts. Annette had been breeding Guinea Hogs for almost thirty years. One day, Annette recovered some well-organized notes she had written in 2005 that detailed her foundation stock and dates they were acquired. That document was a huge help in putting all the pieces together.

I was pleasantly surprised that Becky was willing to take on the challenge of caring for this entire herd until they were evaluated, separated, and observed. She had a small landholding, so this was a huge commitment. The boars and sows had not been separated for a while. When boars and sows are housed together, the boars will take the larger share of food, resulting in sows who are drawn down and need to be nourished back to breeding condition. I knew Becky had a lot of work ahead of her.

Chapter Eleven
The Historic Herds Network

> "Communication networks can be indispensable in placing animals. Some endangered lines of Guinea Hogs have benefited from networks that were able to rescue them after owner situations changed and endangered their survival."
> *D. Phillip Sponenberg, Alison Martin,*
> *and Jeannette Beranger, Managing Breeds*

Organization of the Historic Herds Network

I wrote a letter to the AGHA board about the recent acquisition of rare genetic strains. The board had a closed herd policy that began in 2009. This meant that only offspring of registered hogs could be registered. That policy prevented the Sumrall and Hesters hogs from being entered into the herd book. Becky and I had direct contact with the breeders and had gathered oral history. I had ample documentation that the original board of the AGHA and The Livestock Conservancy were aware of both herds. Both groups had recognized them as potential foundation hogs back in 2005. DNA from the Sumrall boar and Read, Watkins, and Hale bloodlines were documented in the literature. I had confirmed details through direct conversations with Annette Hesters, Gary Sumrall, Don Oberdorfer, Kevin Fall, Shirley Sullivan, Arie McFarlen, and staff at The Livestock Conservancy.

In my letter, I explained the background of the hogs we had obtained and their importance to the health and biodiversity of the herd. In addition, I requested that the board develop a policy for entering herds. I suggested, due to bylaws requirements, that the policy be voted on by the general membership at an annual membership meeting.

Becky and I teamed up in the meantime, staying in touch by phone and email. As my research provided additional information about the Hesters herd, I gave Becky specific questions to ask Annette. Eventually, Annette remembered important details and located her breeding records.

During this time period, Becky and I, along with a handful of other breeders, discussed important topics concerning genetics. We did this via a teleconference book study of the first edition of *Managing Breeds*. There were several pertinent topics that Sponenberg and Bixby outlined that really jumped out at us in this situation. One issue had to do with the conservation of landrace breeds once their environment is altered. "Conservation must therefore result from carefully crafted intentional efforts rather than by the default of isolation that can no longer be realistically achieved," they wrote.

Following that best practice of making breeding decisions by intentional efforts would require observation, documentation, thoughtful breeding practices, and a way for breeders to communicate. Landrace breeds are "historically and biologically important genetic resources that are adapted to difficult environments. They are generally productive with few inputs, having excelled in survival and adaptation." We set an intention to continue adapted traits by monitoring resistance, breeding production, and thriftiness in our herds. We would make selection choices based on those characteristics. We also realized that moving these traits to the next generation involved educating our buyers.

The study group discussed the concept of isolated strains in landrace breeds. Sponenberg and Bixby warned, "The isolated strains… have a high risk of loss or extinction due to dispersal or natural disaster.

The persistence of distinct foundation strains assure that genetic diversity is maintained within the breed. Keeping these separate groups intact assures that distinct lines are present within the breed, so that every animal within the breed has an unrelated outcross in the event that such measures are needed. The usual reason for needing such and outcross would be diminished vigor in a line as a result of inbreeding." This concept of assuring an outcross for every animal was repeated dozens of times throughout the book. It was obviously very important, and something that was addressed in almost all, if not each of our conversations. We realized that some degree of linebreeding is required when breeders want to preserve specific bloodlines. Without close breeding, a linecross would not be possible.

When the precursor to the AGHA was in its planning stages around 2004, the organizers started with three strains and started breeding. They had no pedigrees or oral history to determine how inbred the hogs were initially. They did not know that the Nebraska herd of hogs was composed of direct offspring of the Ohio hogs. They did not have a large group of breeding livestock. So they did what they could at that time—breed and "get the numbers up." Breeders who stayed with only Celesky and Setty hogs had a distinct strain, and some of those are still around. However, most pedigrees today will have a mix, including Biggers along with Celesky/Setty. Some have Baylis bloodlines mixed in, as well. Sullbar Ranch did a good job of linebreeding Baylis line hogs indicated with a VA designation in the name. Kirk Fackrell kept a closed herd at Cascade Meadows. There was, however, a wide focus among many members to breed for low coefficients of inbreeding by always choosing mates that were "not related." That predominant strategy resulted in largely mixed pedigrees. Many breeders prefer this strategy. For the health of the total herd, at least a few breeders need to be linebreeding strategically. Our *Managing Breeds* guidelines drove this point home strongly.

I am not critical of decisions breeders made in those early years. I am very grateful that the breed was preserved and that numbers were up. But Becky and I were in a different space. We had ten years of history and documentation of the breed and genetic lines. I had historical information from the early documented breeders. We had a unique opportunity to move the Sumrall and Hesters genetics forward in a strategic way. I continued to conduct interviews, review documents, and piece together what information I could. It would be another three years before I reached the point where I am today, laying out a detailed and coherent story with few missing pieces.

We had acquired linebred hogs from new strains. We had the *Managing Breeds* text to guide us. We had the Internet, telephone, and Facebook for communication. We had the opportunity to make informed decisions that could benefit the breed rather than only those that provided short term results for our farms. We had a very unique opportunity to employ linebreeding while understanding the traditions of the original breeders. Up to this time, the history had not been documented, and pedigrees had not been widely shared or discussed in regard to breed management.

Becky and I also realized that our Sumrall and Hesters hogs and their genetics could not impact the national herd if they were not accepted into the registry of the national herd. First things first. Then I received additional news.

More Hogs to Recover

On March 2, 2016, I opened another email from Brent Sumrall. His father had passed away a couple months earlier, and the person who was expected to pick up the remaining Sumrall hogs did not come. Brent and Janis had a barn full of meat hogs and some of their best breeding stock. They did not have time to care for them, and several of the sows were bred. They were reaching out for help, so I asked Brent to explain the situation more clearly.

Brent replied, "Hey Mrs. Cathy, I have many of the meat hogs left, and I still have Sadie and the old boar and I do have the papers on the little-boned stock. If you want all the breeding stock, I am going to give them to you. All I want is some of the breeding stock of our old bloodline back one day, but I don't have time for them right now. It may would be a few years or something that is if I decide to get any at a time in the future."

I still felt that Big Boar and Sadie needed to be placed carefully. They had been bred by Arie from Sumrall Bobbie Sue's lineage and had an exemplary body type that I had seen the previous November. I was exhausting my options in re-homing the hogs. The first breeders I reached out to did not respond to my urgent request for help. I wanted the hogs to be carefully placed with someone who understood their significance, but also wanted to assist the Sumrall family as soon as possible. I had a Facebook page devoted to discussing pedigrees, and I had the book study group, so I reached out to those two sources. I received about a dozen responses offering to help, but the hogs deserved someone with experience and commitment. I screened the applicants carefully through email and phone calls, and took careful notes on each candidate. I ultimately chose two women to help me. They had several years of farm experience under their belts, were members of the AGHA, registered their hogs, and were participants in the genetic discussion group with Becky and me. Those women were Donna Dorminey and Deborah Baker.

Not only did these women volunteer to help with the Sumrall hogs, but they were also vital in the recovery project. The three breeders who ultimately worked with me each participated in a phone interview and/or completed a written form so their stories could be shared in this book. I supplemented that information with my personal interactions and observations to create the profiles in this chapter that follow.

Deborah Baker

Deborah Baker, also known as Deb, is owner of Chapel Top Heritage Hope Farm in Tony, Wisconsin. Deb was very dependable about attending the teleconference book study calls. She took notes on things I posted and followed up with emails to me when she had questions about the foundation stock history. She was obviously interested in genetic management of the hogs. In discussions, I sometimes singled her out by asking for her comment or opinion because she tended to listen more than speak. When she did speak up, it was apparent that she did so after careful thought. She was prompt in volunteering to help out with the recovery mission when the opportunity arose.

In addition to pastured pork and carefully selected American Guinea hog breeding stock, Deborah's farm products include some very delicious home made maple syrup, feeder hogs, whole hogs for home butchering or custom processing, chicks, chickens, and eggs. Here is Deb's story written in her own words. (Author's note: This section has been slightly edited for style and consistency.)

"I had been researching and yearning for heritage animals and specifically, endangered hogs. My attention was first drawn to the Gloucestershire Old Spot (GOS) with their place in the farm yard and the looks of the breed. I searched for breeders and at that time there were none near me. My son, Sam Crandall, caught the fever from me and began looking, too. We had become disenchanted with the summer hogs we'd been raising. The feeder hog prices had been going up, while the quality of meat was unsatisfactory. It was pale, tough, and lacking in both juices and nostalgic flavor. Sam began speaking with some Large Black breeders and grew excited about the breed, learning more and finding some very knowledgeable and respected breeders in our area. He acquired a fine breeding pair of Large Blacks.

"I began reading more and more about the American Guinea Hog in my search for an animal that fit some of my purposes in life and needs, and wouldn't compete with my son's growing farm—Northern Marsh Farm, and reputation. As I explored, I joined several online social groups for AGH interests and pursued the breeders' list on the AGHA. In short time I found an ad for shoats from Ricardo and Angelia Silvera's God's Blessing Farm, and bought two gilts from them and my love story with AGH began. In just over a year I have over forty American Guinea hogs in my herd.

"The list of attributes that appeals to me in the American Guinea Hog is long and grows longer with familiarity. The first thing that drew me to them was that they needed saving, nurturing, and preserving. That's my purpose by nature, as a farmer and mother.

"The second draw was their size. Being sixty-three at that time, most of our friends and community were retired or small families. We don't want a freezer full of 500 pounds of pork every year or the half a hog that supplies only half the cuts. The American Guinea Hog puts a usable amount in the freezer with all the cuts and choices.

"I wanted a hog that would be happy out in the field where they belong, a pig that wouldn't sunburn, a pig that would do all the things a pig does naturally—root, feed themselves, procreate successfully, and enjoy mud! We have a lot of mud a lot of the year. The American Guinea Hog has also proven to be exceptionally winter hearty; burrowing through deep snow, nesting in sub-zero weather, and still growing. I found when I raised summer pigs [of other breeds] that once the temperatures fell below 60° F, all their feed went to keeping them warm and growth stopped. Butcher time."

Deb was especially interested in breed conservation and felt that she could help out in some way with the continuation of the breed, even though it was fairly well established. She explained her growing knowledge of the gene pool.

"It has taken me a while to understand the farm lines such as Setty, DNC, Celesky etc., and to place them with herd characteristics. I still feel I am in the learning stages as even the small number of hogs originally saved can become quite convoluted. Then, with the trend to breed for lower COI, the lines were becoming diluted. You can only imagine my excitement when I heard that there were still some regional pockets and old farms with original herds including farm records and hand drawn pedigrees for some.

"Mrs. Cathy Payne was trying to organize homes for some of these animals through an AGH social group. I'm not sure I realized at the time the importance of these hogs, just that they were in bad shape and in need of rescue due to the breeder having terminal cancer and he could no longer care for them. I did not have the honor of meeting or speaking to the owner. The rescue was a passing of hogs, hand to hand, across America following the breeder's death.

"I have read the transcribed interviews between Mrs. Cathy and Mr. Gary Sumrall describing the herd's history. His vernacular in telling the story of his family back through generations of raising the American Guinea Hog was the heart I had been looking for to describe my search for the heritage pig. My own history is steeped in the smells of home-raised pork sizzling in Grandma's warm kitchen, the "pop" and juice of the breakfast sausage, the aroma of coffee and eggs frying in churned sour cream, and butter delicately interlaced with the pork.

"My own wiry Grandpa balancing pails of *slop* across the Minnesota farmyard to outdoor pigs that looked nothing like today's "confinement pink" pigs and coming in for that breakfast at the break of dawn. I want my grandkids to know those smells and memories. Mr. Sumrall's interview brought those memories alive with a delightfully southern touch. Mr. Sumrall's son Brent (Buck) Sumrall is aware that I have three of his father's hogs in my care. I would be happy to supply him with old stock bloodlines in the future when he is ready to breed again.

"I have a great respect for these hogs. They came to us in terrible shape, worm-ridden and starving. They had been contained in a dark barn while help was sought. Their coats were thin and hooves long. I believe it is mostly due to the stamina of the American Guinea Hog breed that they hadn't perished. They came to me scared and in need of help. It is my responsibility to sustain them as purely as possible."

Deb continued, [The Sumrall hogs] are a rediscovered old herd that hasn't been routinely linecrossed. Old farmers tended to have sows and a herd boar. They bred them and the weak or deformed didn't survive. Over generations this created a strong dependable herd of one type of hog. That hog was their farm hog. I'm very excited about the discovery of this farm line and their history.

"I believe that the diverse farm herds of American Guinea Hogs as they have developed over generations in the seclusion of old family farms should be kept as they are, and valued for their uniqueness. I have Setty bloodline hogs that are longtime Northern hogs. They have characteristics that protect them from the harsh cold and resulting frost bite, smaller ears, shorter noses, and thicker winter coats. The Sumrall hogs are a southern pig with very long noses for rooting deeply and large hairy ears to expel heat and discourage biting flies. These characteristics would help them survive extreme heat, drought, and year-long pests.

"The Sumrall Sow, Big Momma, has magnificently large, upright furry ears, a long face and snout, clear dark noticeable eyes, and a single curly tail with a tuft. As she fills in, I can see she is larger than my Setty hogs and, while different, is definitely American Guinea Hog conformity. They are good hogs. They can be handled, are friendly, curious, and active. They move well. The sow is a good mother after an unassisted farrowing, and has a bountiful milk supply. She is gaining condition, albeit nursing. The babies demonstrate the same good qualities, including a higher teat count than the Settys. They seem bright and have begun eating grains and grasses early on."

As of 2019, Deb has three genetic lines on her farm—Baylis, Setty, and Sumrall. Her goal is to use strategies of both linebreeding and linecrossing. She culls late and evaluates each generation. In addition, she continuously monitors her stock. She pays attention to conformity to AGHA breed standards, including a preference to solid black color. She breeds for the continuation of traits more specific to her three genetic lines.

Deb is concerned about preventing detrimental congenital traits. Piglets of any breed can be born with hernias, cryptorchidism, weak pasterns, humpback, and other weaknesses. There is another trait that she culls for that I need to explain for clarification. Many Guinea Hogs display a tendency to flop over for belly rubs. When approached and scratched on the belly, these hogs will flop over and relax. They may go into a "zone" for a minute or two, quite relaxed and still, until the scratching session ends. The trait tends to be inherited and is more prominent in certain lines. Guinea Hog breeders often refer to these as "floppy pigs." Many breeders select in favor of those most likely to be "floppy" when choosing breeding stock, because it is considered a sign of a calm and gentle temperament.

Deb feels that the floppy trait is "not cute" and could be "detrimental to the pig in natural survival." She thinks of it more like *caprine myotonia*, or fainting goats. Myotonia is defined by extreme muscle spasms. In affected goats, it can be triggered by sudden, loud noises. Deb reported to me that she has seen videos of Guinea Hogs that "freeze," with breeders lining up three hogs in a row that are still and remain still after touching them. Deb considers that observed stiffness to be a neurological problem rather than a state of relaxation. On the other hand, she does raise hogs that will resume standing or moving as soon as a belly rub session is over but do not remain in a "frozen" position.

Deb preserves her three lines by linebreeding related hogs. She told me that linecrossing will be used "only when necessary to eliminate surfacing inbreeding deficiency, then bringing the best back into that herd." Hogs selected from linecrossing will also be made available to customers requesting low inbreeding. In all cases, she strives for high levels of quality and to produce an excellent type in her breeding stock.

Deb chooses to sell breeding stock only to customers located at least one hundred miles from her farm. This helps avoid having too many genetically similar hogs in one area. She believes that an overabundance in one area will cause a decrease in interest and will lower prices. She has respect of the hogs in her herd, and thinks that the breed should be a cherished part of a farm's livestock. She has the highest regard for the Guinea Hogs' longevity and independence. As a final thought, she stated that "We should be only protectors of an already perfect breed."

Donna Dorminey

Donna contributed in a major way to the Genetic Recovery project, which will be detailed in Chapter Twelve. Her role in the project was enormous and the actions she took are likely to make a positive difference in the overall national herd for many years to come.

Donna lives and farms in Upton, Kentucky. She is an experienced horsewoman and show judge, and her farm is named after a beloved horse, Solomon's Wisdom. Donna is a retired lieutenant colonel in the U.S. Army. She is currently employed full-time as an operations research and systems analyst in the Army. Schooled at Cornell University, she has a sharp mind and a tireless work ethic. She is skilled in statistics and knows her way around a computer. She is working on her doctorate degree while running Solomon's Wisdom Farm, tracking her hogs' genetics, tending her horses, and working full-time. Her work requires her to travel extensively.

Donna's mother served in the military, and her daughters, Katie and Samantha, are both second lieutenants. Between them, they represent three generations of auxiliary members. Hers is truly a patriotic family. Donna seems to have a great relationship with her parents and daughters.

Donna's foray into Guinea Hogs started out "as a little bit of a joke with my father," she told me in a phone conversation in June 2016. "My daughters and I were looking at increasing animals, and we had Pygmy goats. We were not fond of them. We came across, actually, a Craigslist ad. Not too far away. About an hour away. And someone was retiring out of raising Guinea Hogs. So we went down there and looked at them, and we bought our first registered breeding pair. We kept just the one pair because it was only my two high school daughters and I here."

Donna's first duo, Rudy and Sasha, represent lines from Celesky [Setty] and Baylis. "Of course, I didn't know anything about pedigrees or lines or anything like that at the time. So that really had no meaning to us at that point. We liked the look of them and their personalities, and they were very friendly. So we thought, 'Let's give it a shot.' They are very personable and easy to care for. We absolutely love them. We decided to stick with just one breeding pair because we didn't want to overload ourselves. We have horses and chickens.

"We got into American Guinea Hogs because of that homestead characteristic. They are small, friendly, and easy to manage. They are not overwhelming. We are not looking to make money. We just love friendly hogs that we keep on the farm and don't have to worry about safety with anybody. We let them have a litter to two litters a year. It was all about just putting pork in the freezer and some for family and friends and that type of thing. And selling a breeder every once in a while if I had a really nice pig. I have sold a few. Probably less than ten over the last few years because we are very selective about who goes out as a breeder. Our purpose was not to sell breeding hogs or anything like that. It is really the homestead and providing meat for our family and friends.

"And that is the way we have gone about raising them for the last six years. Until this year." I asked Donna to tell me more about what she looks for when selecting breeding stock.

"My background is in horse and horse judging. I am an engineer, so I look for form to function. I am looking for correct structure, and I am looking especially for legs that can bear the weight of the Guinea Hog. That sloped pastern, but not where the dew claws are on the ground. And not upright where there will be a lot of joint wear and tear. I'm looking for the gently sloped pastern for the movement of the hogs. I look for good bone structure.

"Like a horse, I look for nice, thick cannon bones. That is the bone between the hip and the knee. That is what is going to hold this animal up for its whole life. It is similar to a horse. I look from the front and the back and well-set legs. So when they stand, they have a good stance up front. The front feet are not together. I like when their feet are straight. Pointing straight forward. A little bit of toed-out is okay, I think. But I don't like a lot of toed-in. That will put stress on the joints over time. In the back, I take a little bit of what we call cow hocks, where they point in. A little bit of that. But if they are too cow-hocked, where the hocks are too close together, that is not something that I like. I like a nice spread in the legs. And then I look for the flat top line. I like the balance between the hind quarters and the shoulders. I don't like one to be massively bigger than the other. People talk about the rectangle look from the top.

"If something is going to be bigger, I prefer the shoulders to be a little wider than the hips than the opposite way around. And then for headset, a lot of that is preference. I prefer the shorter snout to the wider snout. When I look at the head, I prefer slightly bushier ears. And I look for the personality. That is a gut feeling type of thing. It is really hard to describe."

Becky Mahoney

Becky and her husband Patrick own Joyful Noise Home-N-Stead in Macy, Indiana. Becky is a loving mother, daughter, and wife as well as a homesteader who can do just about anything that needs to be done. Becky discovered the Guinea Hog breed in late 2013 or early 2014. She was drawn to both the Guinea Hog and the Mulefoot breeds but settled on Guinea Hogs after speaking with other breeders. She purchased her first hogs in May 2014, and wrote about them for me as follows. (Author's note: This section is slightly edited for consistency and style.)

"My first two [Guinea Hogs] were unregistered, purebred, 'bred' gilts. I figured that starting with unregistered gilts would be a low-cost introduction and give me a chance to have farrowing experience before the registered stock birthed. It did not work out that way, as neither one turned out to be pregnant. Two weeks after getting the unregistered gilts, I found a registered breeding pair in southern Indiana. A couple of months later, I picked up yet another breeding pair in Kentucky—the gilt for me, and a boar for a friend. In the spring of 2015, I bought a herd in Georgia from a herd dispersal. In the fall I picked up an older boar in Oklahoma. Then, in December of 2015, I bought the Hesters herd. I have learned much in evaluating the different animals. I culled hard based on soundness, structure, and temperament as well as mothering ability and quality of piglets."

Explaining her interest in breed conservation, Becky states, "I have already raised heritage chickens and turkeys. I appreciate the full flavor taste and texture of the meat of a slower growing carcass. The modern, fast-grown meats are flavorless and have a soft, mushy texture. I also like a challenge, and I do not follow the status quo, so working with an endangered animal fit the bill."

Becky developed a relationship with Annette Hesters. She remarked, "I think we made a connection and a friendship. We talked about her herd, her family, her struggles, her enjoyment of antiques and history, and, of course, the history of her AGH herd. I bought the entire herd with the understanding that the following spring she would get back one of her little ones and one from my herd, thus a young breeding pair. Annette is a sweet lady that loves her Guinea Hogs and many smaller animals. It was a difficult decision for her to part with them, but that decision was made easier knowing that she would get a young breeding pair the following spring. Even though she enjoys the hogs, she did not spend time with them hands-on such as petting or scratching them."

Becky continued, "I am responsible for any animal in my care. I give them the best nutrition that I can, work with animals to tolerate humans, and breed them responsibly to improve the herd dynamics. I cull hard. I breed carefully."

She elaborated on the Hesters herd, "My understanding is that two sows and three young carry genetics from the Read line. This has been a closed herd for over thirteen years, and has been in existence since the late 1980s. The AGH population of registered stock originated with a very limited gene pool. It is of utmost importance that the pool be nurtured and increased.

"The Hesters hogs bred from [Brothers] mixed [Setty/Celesky/Biggers] stock from 2005 are small. The mature boars are likely not much over 200 pounds. The sows are even smaller. The two sows with Read genetics have a bit more skeletal structure, but are also what I would consider small-boned. They all meet the description of AGH, though some have more erect ears, and some have ears that tend to shade the eyes. All are solid black, with a curled tail, upright ears, and dense hair. Hair can be either straight or curled.

"From here, we keep plugging on. Not all will produce quality stock. Those of lesser quality will taste really good! I think that it is important to learn to market our hogs, getting others interested for homestead and restaurant use. Now I am working on getting a foot in the door with a farm-to-fork restaurant in northern Indiana."

In 2018, I interviewed Becky to find out what she had learned about the Hesters herd over time. She told me that even within a single litter, she had observed a variety of structure, shape, hair, and faces in the herd. Some had more forward-facing and shaggy ears, while one she named Esmerelda had a very erect ear carriage.

"All Hesters have been small," she told me. "Annette kept small animals." Hesters Charles at age two was 150 pounds, and Hesters Ajax at age four was 160 pounds. Sows are even smaller, about 125 pounds at age two. Hesters Fiona, age three, is no more than 130 pounds. "I'm really appreciating their small size, " she stated.

Becky got very enthusiastic when describing her herd. "I think they're beautiful hogs! They're smaller. They just have a really nice shape. They're not getting fat really super fast. They're not hard to keep on the leaner side. Fiona did really well when she had litters. She didn't nurse down skinny. All of my sows have had cone-shaped udders when they're nursing."

She told me about piglets born in 2017 that were still growing out in 2018 to be evaluated for temperament. "Some piglets from last year are very amicable to being scratched, some that are coming around, and some that still don't want to be touched. And that's out of one litter. So we're still working on the temperament genetics."

In regard to the breeding pair, Hesters Fiona and Hesters Charles, Becky said that they "are easygoing. You can scratch on them, and they flop over." Becky has not been at all hesitant about culling for bad temperament. This is a hallmark of the breed, and aggressive hogs must be eliminated as breeders. She has "honed in to Charles and Fiona to move the Hesters genetics forward to the next generation."

Becky shared her thoughts on selection characteristics. "When I was at The Livestock Conservancy Conference, I went to a session on genetics. They were talking about a breed of sheep that had been secluded and wild for decades. They were very hardy. And when they increased the size to compete in shows, they lost [much] of their hardiness. I was thinking about the Guinea Hogs. I know that there are people [selecting] for bigger and faster-growing hogs. They're trying to get them to more of a commercial type, when that's not their place. Their place is hardiness, and genetic diversity. This is an animal that will survive, and smaller is not bad. The mindset of, 'Let's get them bigger, faster'— that's a commercial mindset."

Becky's breeding goals include preserving and disseminating the Hesters line of breeding stock for another generation to enjoy. She wants to use the strategy recommended by Phil Sponenberg, which is to alternate linebreeding with linecrossing. For this purpose she employs two boars in her small herd. She selects breeding stock that has typical gentle temperament, exemplary mothering ability, and sound structure. "I observe how they grow out," she told me. "Im not intent on getting two litters a year. That would just get me overrun with pigs again."

I appreciate everything Becky has done. Without this intervention and recovery of the documents from Annette, the knowledge and the genetics of the Hesters herd would have been lost to history, and I never would have tracked down the Hale or the Read families.

Cathy R. Payne

I raised Guinea hogs in Elberton, Georgia from late 2013 to mid-2018. I first heard about the breed in 2013 from one of my heritage rabbit livestock customers. She had just picked up an unregistered trio and raved about them. I began to do some research, but as I explained in Chapter Ten, there was just scant information out there to satisfy me. That was when I began delving deeper. Once I had some basic genetic information under my belt, I started to locate Guinea Hogs near me. I didn't know much about evaluating the hogs, and in one or more instances, neither did the sellers. But I observed the hogs closely, collected data, asked questions, and learned as I went along.

My farm, Broad River Pastures, had a total of eleven acres. Only eight of those were fenced because we kept a portion of our land for conservation of wildlife and a riparian buffer on the edge of a creek that fed the Broad River. My husband and I "retired" there in 2010. After about a year, he started his own service business and I started the farm business. I raised Gulf Coast Native sheep, American Blue and White meat rabbits, black Silver Fox meat rabbits, heritage chicken breeds for eggs, and Khaki Campbell ducks for eggs. We specialized in breeding stock, meat, eggs, and high quality herbs including comfrey, tulsi, horseradish, catnip, turmeric, and ginger. I started an internship program to train future farmers.

I started raising Guinea Hogs with a herd of three gilts and two bred gilts. Like Becky, I didn't have much luck with the bred gilts. One of them crushed her first litter, and her sister abandoned hers. Both sows became freezer meat, the first one after a couple litters, and the second one as soon as her milk dried up. My efforts toward contributing to and promoting the Guinea Hog breed included:

- Membership in the AGHA and The Livestock Conservancy
- Volunteering for the AGHA Board of Directors
- Writing newsletter articles for the AGHA and The Livestock Conservancy
- Sharing my knowledge with customers
- Giving homestead tours
- Doing chef tours
- Interviewing for podcasts and newspapers
- Managing a Facebook page for breeding discussions (Registered American Guinea Hog Genetic Discussions)
- Developing a tool to help breeders select breeding stock (AGH Selection Matrix)

- Presentation with Donna Dorminey for The Livestock Conservancy's Annual Meeting
- Facilitating a teleconference book study on the first edition of *Managing Breeds*
- Facilitating an online Google Groups discussion of the second edition of *Managing Breeds*
- Maintaining an educational website at www.guineahogbooks.com with a pedigree gallery

My Guinea Hog herd was just amazing. I learned so much from observing them, learning their habits, touching them, and talking to them. If you've never snuggled up to a warm, hairy Guinea Hog, scratched its belly, and patted its rump, you may want to add that to your bucket list. It is truly priceless. My personal favorite pastime is massaging the swollen belly of a favorite sow a day or two before she farrows, talking to her gently, then watching her waddle to her wading pool and lay down for a good soak. But I guess even more than that I love the next morning when I pick up a newborn piglet and hold it up to my cheek. And then there is watching a squirming litter of eight or more piglets scrambling for the sow's teat, making the most of it while mama lays down for them until she closes the milk bar. Every moment is special with these hogs.

  The people I have met along the way have greatly enriched my life in ways that I cannot describe. Once I added the Sumrall hogs to my herd, I gained even more admiration for this wonderful breed. The genetics they added greatly improved what I already thought was a vigorous, healthy herd. In order to document that even with inbreeding, there were not deleterious genes in the Sumrall bloodline, I held back boars and bred them to young gilts to produce three generations over a short time period. I had five different boars on my small farm. Each hog had unique characteristics and personalities. I loved each one, called them by name, and sang to them.

I mourned my loss for months after making the decision to sell our farm and place the hogs with other competent breeders. However, I made the best choice for my family and health. I'm not sure if this book would ever have been completed if I were still on the farm with so much to do outdoors each day, breaking up my writing routine. This is my gift to current and future breeders. Knowing where we are with the breed will help us plan and do what needs to be done to move them into the future. It is a crucial time for making informed decisions.

Another Trip to Mississippi

Now you know the major players and their personal motivations to preserve biodiversity and expand the genetic field for the Guinea Hogs. In the first week of March 2016, I put out a plea for help with a dispersal of the Sumrall's unregistered but rare hogs. Donna Dorminey offered to help on March 4th, and Deborah responded to me one day later. I admired them both for their interest in breeding strategies and the history of the hogs. I knew they owned copies of *Managing Breeds for a Secure Future*. That reference book and the second edition, when it was later released, forged our guidelines moving forward. Becky, of course, had been working with me since November. I had my team!

Donna made the drive to Laurel, Mississippi on April 16, 2016 to pick up Big Boar (later determined to be Maveric Hansel) and Maveric "Sadie" for her own farm, and an older gilt for Deb. When Donna arrived, Brent and Janis Sumrall appealed to her to take the remaining breeders. So in addition to the preceding hogs, she got a sow named Big Momma, two boars under the age of two, four gilts between one and two years old, and one gilt less than a year old. One young boar and Big Momma would eventually go to Deb Baker along with the gilt.

THE HISTORIC HERDS NETWORK

The breeding hogs that now remained on the Sumrall farm included only one boar. He was a short, long, little-boned hog with no Maveric or Sumrall bloodlines. He was descended from registered small-boned Shirley family-owned stock in Mississippi. We would later learn that the boar's grandparents were from Carolina Heritage Farm. He had the characteristically long nose and short stature common in Gra' Moore's herd. The Sumrall family still had about thirty barrows and meat-quality gilts that needed buyers. Of the ten Donna picked up, she would keep seven. Three of them would go to Chapel Top Heritage Home Farm in Wisconsin.

Donna's trip took place about six weeks after Brent's second call for help and five months after the Sumrall family first began their herd dispersal. Remarkably, Donna had been raising only a single duo for several years to avoid overextending herself. Now she was not only adding seven additional hogs to her breeding stock, but also getting involved in a genetic recovery project. Each of us in the communication network knew that the offspring of our new hogs might never be registered by the AGHA. We took that risk for the sake of the breed. I held out the hope that even if the current board of the AGHA rejected the admission of the stock, that at some future date the hogs we raised, marked, and documented could be included. The diversity was important for the breed. All four of us were committed to taking on this risk, subjecting ourselves to scrutiny, and expanding our herds, our infrastructure, and our budgets dramatically. In Donna's case, she told me she had been following my research and was "eagerly awaiting the publishing of your books."

Donna continued, "And when I saw your post wondering if anyone could take in some of the Sumrall hogs, I kind of jumped at that opportunity. We had a discussion here, because we knew that it would change our way of doing things here on our farm. We had just stuck with our homestead up to now. We had a discussion here at home before I called you back and said, 'Hey, we are really interested.'

And we really think it is neat. That is one reason that we really like the Guinea Hog. It is a tenacious little breed. To be involved in bringing it back….When I first got involved, it was because you needed somebody to go out and get them. And we looked out here and said, 'I have thirty-two acres of pasture. It would be very easy to take care of these hogs, even if it was temporary.' We could get them out of the situation that they were in and get them safe and taken care of for whatever was decided after that. That was our original plan. Just to be part of saving this great breed. And then we got them here, and they are great.

"We already had a great breeding pair, Rudy and Sasha. We love their personalities. And just to think about taking part in conserving them. What really did it for me was to go down and talk to Brent, and talking about how important they were to his dad. And seeing the tears in his mother's eyes as we were getting ready to pull out with them. Knowing this is a part of her husband that she was losing, after losing him. It was really hard for them. Their hearts and souls were really in it. We told Brent, 'If you are ever in the position to get them back, there will always be stock to come back down here for you.'

And that is our commitment to him. And it is a major commitment to him. A part of our commitment is not only will there be stock here, but stock that has been selected for the characteristics that his daddy selected them for. We will continue the Sumrall line by selecting for the Sumrall traits that were important to them. We had a discussion with him on what his daddy looked for in the hogs, and what he prized. What his daddy did."

This was vital information that Donna had the foresight to gather from Brent and Janis. When Pam and I visited five months previously, Brent and Janis stood back and let Gary do all the talking. While Gary was having a "good day," there were bottles of medicine on a nearby shelf, and he could not sit upright on the sofa. I know the interview tired him out. He did not provide much detail about selection traits, or what he looked for in a breeder, even when prompted. He primarily emphasized that he was looking for "stock like Daddy had."

With the last of Gary's breeding stock leaving the farm, Brent was highly motivated to share information with Donna. She was in a position to preserve the family legacy. The genetics of the hogs could potentially return to him one day. He remembered that his grandfather had difficulty retaining the right type, according to Gary's memories in 1961. And in 2012 when the family located Guinea Hogs in Mississippi, they were inferior, by Gary's standards. Brent provided a verbal picture to Donna as the hogs were being loaded.

Brent pointed to "Big Boar" (Maveric Hansel) and told Donna, "That is Daddy's ideal." Donna asked Brent what his daddy really liked about him. Donna repeated that information to me. "And we sat and talked for about five minutes about Big Boar. We talked about the slightly forward-facing little bit of shaggy ear with a little flip at the end of them. He talked about the shorter snout with a little bit of upturn at the end. 'Daddy didn't like long snouts, and he didn't tolerate anything but black,' Brent told me. 'If there were any red highlights or anything, he culled.'

Donna continued to tell me that Brent discussed how wide Big Boar's shoulders were, as well as the hips. "And about the bone quality and how you can't even get your hand around the cannon bone in Big Boar. The bone is so substantial that you can't fit your hand around there. The cannon bone is the one that goes from the fetlock and, you know, from the pastern. It is the one that goes between the fetlock and the pastern. That, and the knee. In horses, it is called the cannon bone. I am assuming that it is probably called the same thing in a pig. (Author's note: It appears to be called shank, foreleg, or hock, on charts I have located) I don't really know for sure. That is what we call it because of our analogy. And he talked about the wide spread, the four corners, and able to take the weight. And we talked about them being a little shorter and really broad."

I responded, "I know that [Gary] didn't want to sell everything. I mean, he had only had the old stock back for two years before [he] died. And he had a lot of culls in [his barn]."

"It makes sense," Donna replied. "Because just from talking to Brent, it was clear that Gary Sumrall was very selective on what he would keep to breed. Everything else was a cull."

"I don't know if you have ever heard the story about when Gary's father died," I mentioned to Donna. "When Gary's father died, basically on his death bed, one of the last things that he said was, 'When I am gone, Gary, please take care of my pigs.' So that was a promise Brent's father had made. And he almost lost them when he had that contaminated feed. They had already lost their old boar and couldn't find a replacement for him. Then they only had three old sows including Bobby Sue. The fact that we have these hogs at all is because Arie took them. Only Bobby Sue ever farrowed again. Arie kept [the offspring] boar, Maveric Charles Sm3. We would not have these hogs at all, if not for her recovery project. It was a close call."

"It is. It is amazing," Donna replied. And I think that is why we need to be as selective as Gary was." Donna told me a bit about her selection process. "I like to wait as long as I can, to really know. I think you can start to see a lot of their personality and stuff come out. But I have only had them for almost six years now. And I guess I don't have the confidence in myself to see it in young piglets or shoats. I am starting to notice, and I think I will get better at it. But I like to wait and make sure."

Donna's pick up coincided with a surprising bonus that would be invaluable to the recovery project. Janis and Brent, in the process of going through Gary's things, discovered pedigrees! There were registration papers for the gilts and boar that Gary purchased from Dean and Rhonda Shirley of Mississippi in 2012. These were the little-boned hogs that leaned toward red highlights and red tinged ear tips. There were also Maveric pedigrees from Arie linked to "Big Boar" and

"Sadie's" ear tags. Between the ear tags, pedigrees, and DNA testing, we would eventually sort out the relationships between the Sumrall hogs. The pursuit of historical information was resulting in one remarkable discovery after another. And there was even more to come.

Sumrall Feeder Hogs

All that was left now was to help the Sumrall family out by getting a buyer for the feeder hogs and the little-boned boar. Gary had grown fond of the odd-looking boar, with its short legs, proportionally extra-long body, giant tusks, and long nose. He was the only boar Gary had for the years before his family stock returned from Maveric. In order to avoid excessive inbreeding, he continued to use him on the Maveric bred stock. He even planned on keeping him if his health had improved. The feeders were culls, both barrows and gilts that did not live up to Gary's high breeding standards. The family was hoping to get enough money to recoup the cost of the one or two years' worth of feed it took to grow out those feeders.

In November 2015, I got a call from Marcos Fernandez, a chef opening a new restaurant in Lakeland, Florida. Named Nineteen61, it had opened in December, and he wanted to serve primarily Guinea Hog pork. The son of Cuban immigrants, Marcos wanted high quality pastured pork, because pork is important in Cuban's culinary culture. He had read that the Guinea Hogs once thrived in Florida. I told Marcos that I would be unable to provide meat for him, but I could help him find a local breeder if he was unable to do so. I thought that maybe he had someone who could get the Sumrall hogs, fatten them up, and sell to him.

I had contacted one farmer who was interested in picking up the Sumrall hogs for his meat business. However, at the last minute something came up that prevented him from following through. I checked with Marcos, who was interested, and asked him who had been growing his pork. He directed me to Bob and Karen King of Mt. Citra Farm in Citra, Florida. They hesitated to pick up the hogs at first, because the price was not low enough to make a long trip worthwhile. Eventually, an agreement was made with the Sumralls, but the Kings could not get off the farm to make the pickup.

They hired a subcontractor to transport the hogs from Mississippi back to Florida. When the person arrived for the hogs, they had lost condition. Brent convinced the driver to take the little-boned boar and the rest of the animals not purchased by Mt. Citra Farm.

In the end, the Sumralls were relieved of some great pressure and put some closure to that chapter of their lives. In addition, Marcos had a nice influx of Guinea Hog pork during his first full year of operation. The Kings had a short period of tending and investing in the hogs until their body condition returned and they could go to the processor, and the subcontractor had a new boar and more. This brought a closing for the Sumrall family and doors opening for others.

Chapter Twelve
The Genetic Recovery Project

"Landrace conservation has many inherent challenges, not the least of which is the very choice of which specific individuals should be included in any landrace….Landraces are by their very nature more variable than are other classes of breeds. Landraces are characterized by a consistency of biological and adaptation traits, and not necessarily by uniformity of physical appearance….They are historically and biologically important genetic resources that are adapted to difficult environments."

"For a closed herd book to be valid in a genetic sense, it should include a vast majority of the breed. It is all too common for recently structured landrace herd books to close before a reasonably high proportion of the animals that are actually members of the landrace have been included. Closing the books restricts the number of breeders and animals within the population recognized as valid by the registry….This is short-sighted when considering the importance of the fate of breeds as genetic resources. With few exceptions, closed herd books are inappropriate for landrace or local breeds."
Dr. Phillip Sponenberg, Alison Martin,
and Jeannette Beranger, Managing Breeds

Getting Started

Becky, Deb, Donna and I spent 2016 getting to know our herds, getting our network organized, and periodically communicating via email and telephone. The process of opening the herd book would not be approved for several months after all the hogs were in our care, and then the approval of our herds would take another year. We started to meet monthly to discuss issues and to support each other. The concepts and strategies in *Managing Breeds* became our touchstone during this process. It argued for inclusiveness in a landrace breed, selecting for hardiness, preserving unique strains, using communication networks, and more. We cycled back to it frequently.

While we waited for AGHA decisions, several sows bred at the Sumrall farm were showing signs of pregnancy. Sumrall MS Swanee Rose, bred on my farm after quarantine, had farrowed a lovely litter of six. Life cycles were proceeding. Donna arranged for three Sumrall hogs to go to Deb in Wisconsin. Becky managed the Hesters herd. We settled in and got to know our new stock.

As litters arrived, we evaluated their outcomes. It was important for us to document consistency from generation to generation. At Broad River Pastures, I was linebreeding as soon as my boar reached breeding age so I could get three generations on the ground to demonstrate the breed characteristics expressed. A fourth generation, my boar's dam, was in Wisconsin with Deb. If we had seen any surprises, such as striped piglets, severe deformities, or reduced litter size, those would have been significant signs that our hogs should not be entered into the registry. With our linebreeding strategy, any weak genes or signs of crossbreeding would be accentuated and come out in the phenotype.

But Wait, There's More!
Donna Discovers the Maveric Herd

Then—more synchronicity. Donna tells the story. "I was really just bored at work. I was waiting for a computer process to finish. I was just looking through Facebook ads [and saw these unregistered hogs for sale]. And it was just a few days after we had picked up the pigs from the Sumralls, and I had seen the pedigree down there. Brent had it up on the hood of the car and was talking about Maveric, and then you mentioned it. And so I really heard the story of the Sumrall hogs and stuff."

Donna continued, "When I glanced through the Facebook posting and noticed that they are not registered, so nobody was really interested. Then he [the seller] came back and said he knew the pedigree, and that perked up my ears. Unregistered hogs don't usually have them. If we had not had that discussion, and if I had not been involved in the Sumrall hogs, I would have gone right past that.

"So, right there on the spot I stopped what I was doing and sent him a message. I said, 'Hey, can you check the pedigrees and see if there are any Sumrall hogs anywhere in the pedigree? And sure enough. It came back, and there was! And how exciting is that? That was absolutely incredible!"

She continued, "And then, I am looking at my boyfriend and saying, 'Hey, I know we just brought back ten from Mississippi, but I want six more.' "

At this point I said, "He's a good one. You'd better hang on to him."

"Yes, he is a trooper," Donna agreed. He said, 'So when are we going to get them?' April 16th is when we went down to the Sumrall's and brought those [first] ten back. And then two weeks later on May first, we drove down to Tennessee and brought the others back. I think it is truly historic. When you sit and look, and the more

I learn, the more fascinated I become. And you sit here and think there were eight herds that were known, but only four of them made it into the registry. And we have three of the four lost ones. And not only do we have — she had the last blood from those three or four. It is like there are no others that anybody is aware of. But to think, this is an opportunity to save not only these lines themselves, but to save the breed over the long term." ." (Author's note: At the time of this conversation, Annette had not yet recovered her breeding notes about the Hale and Watkins sources in her herd. We had, in fact, recovered *all* the known bloodlines from the Maveric, Hesters, and Sumrall acquisitions. My research later uncovered herds from the 1990s that have been lost without a trace, but we recovered all of the unregistered bloodlines known by the organizers of the AGHA in 2005.)

"Yes," I agreed. "Because the more biodiversity in the breed, the larger the gene pool, and the healthier it will be."

"Exactly!" Donna replied.

What Donna had found, it turns out, was the person who purchased the remnants of the Maveric Heritage Ranch herd back in 2014. This is the person Rico Silvera and Gary Sumrall had wondered about. The pedigrees Donna had on her farm now included Brown's Blue Boy from Billy Frank Brown's former herd, more descendants from Sumrall Bobbie Sue, and descendants of the Marcia Read's bloodlines out of Iggy and Ziggy from the Hesters herd prior to 2005. In the network, on four farms located in four states, we now had management of all of the known lost herds.

"But I also think," I said, "from the book study that it is really important that some people preserve the lines and do linebreeding and that some do linecrossing. We need some lines fairly close together so we have the lines."

"It takes a mix of breeders," Donna agreed. "A few of us had this discussion before. You have to have the people that will preserve the lines. That is kind of where I see our role right now. Up here, and

THE GENETIC RECOVERY PROJECT

my intention moving forward is to save the Sumrall line. We have not made the final decision, but we have basically closed up the Sumrall herd, and we will not bring in any new blood. And we will perpetuate the line. Then hogs will leave from here to go out and out cross and inject new blood into other people's herds. But we're not going to dilute this herd any more. We will bring back the Sumrall characteristics and then it will stay a closed herd. We will breed this herd unless a time comes that we have to inject."

"And that is the way the old-time breeders did it," I agreed. "They didn't have access to other hogs, especially as they became more and more rare. Gary was looking for hogs to inject into his bloodline, but I think it took him twenty or thirty years to find somebody. I mean, they were just using the old hogs for a long time."

"Right," Donna said. I think with the diversity that we have right now, with these five different hogs from four different litters, we haven't worked the figures yet, but my plan is to work out a little plan and send it to Dr. Sponenberg and see what he thinks. But I think we can perpetuate at least four, five, or six generations here without really being too inbred because these five hogs come from four different litters. So there is only one set that are full brother and sister. And of course, those are from different litters, but still have the same boar and sow. But the others are half. They all have just Charles in common. I really think that because of that, we will be able to come up with a smart breeding plan that will dilute out some of the other and concentrate the coefficient of relation or inbreeding, which way you want to calculate it. It will be focused more on the Charles side because that is what they all have in common. And therefore Bobbie Sue."

"That is excellent," I stated. "And I think that is really important. What I would like us to be able to do as a network is keep track of the people who are continuing that type of linebreeding. So, you know, like Becky is working with the Hesters line and you are working with the Sumrall line. Deb and I will work with the Sumrall line, too, but we each only have two or three hogs to work with, and you have a dozen."

"And there is a little bit of Hesters in my line, too," Donna told me. "And they might be grandparent or great-grandparent, and we need to look at the pedigrees and see if one of those piglets would be good to send to Becky. And I don't know if it is too far back compared to what she is working with or not. So I would love to hear from Becky what characteristics she is looking for, for selection. If I have a piglet out of one of these sows, one of these that does have a tracking down to Hesters, if I had one that has Hesters characteristics, rather than culling it, if it fits the characteristics of a Hesters hog, I would rather send it to Becky.

"So if we can get the characteristics that each of us is trying to perpetuate, because a lot of this Maveric stock does have some of all of them, it would be great to say this doesn't show the Sumrall characteristics, but it has more of the Hesters and that way, it would help them bring back those characteristics. That is what I would like to do. Because Maveric was the mixing pool and the dumping spot. So it seems that the Maveric source will be the key to get back the original characteristics again."

I said, "This is an important time. A very important time. So, yeah, I didn't expect to be smack dab in the middle of it. But I had been trying to reach old-time breeders, and I got a lead on how to get ahold of Arie and Gary. I was in contact with Brent, and you know, it was every month or two I would send them an email. And sometimes I would catch Gary for five minutes on the phone. And he was too busy to talk, and we would go back and forth like that. And then, after about a year, I got an email from Brent saying, 'Daddy is sick, and he wants to you have pick of the litter of his hogs.' So, you know, it was just like really, out of nothing. And it is kind of like you glancing at Facebook in the middle of the day and just seeing Maveric unregistered hogs. And in some way, I think we are all meant to come together at this particular point in time and know just enough to appreciate what we found and know how to keep the lines so every animal has an outcross, like it says in the book [*Managing Breeds*].

"Well, you know," Donna said. "They are incredible hogs. They are important because of the genetics that they carry. But even if somebody told me today these hogs can never be in the registry, I wouldn't care. These are awesome hogs. They have great personalities, and they are a joy to be around. So it would be okay. I guess I will have unregistered hogs because I would carry on. Because they are just wonderful."

"They are," I agreed. "And I think it will happen eventually. It might take a while. So that is why your role of playing registrar and keeping up with all this is really essential. Because we have litters being born. And how many generations? Do we have four or five generations alive and saved right now? We have Big Boar, Big Boar's daughter, and his granddaughters?" The conversation continued for quite a while more, while we compared notes on phenotypes and who was due to farrow. On all four farms, plans were developing as we bonded with our new herds.

The Historic Herds Network

Bit by bit, our network organized and loosely formalized in order to steer us in a good direction. We called ourselves Historic Herds Network (HHN). Using the *Managing Breeds* guidelines, I proposed a motto, mission statement, and code of ethics that was accepted by the group. This gave us some focus and definition.

<u>Our Motto</u>: "Working Together for Genetic Diversity and Preservation"

Mission Statement: The Historic Herds Network is a group of Guinea hog breeders and conservators who track pedigrees, study conservation practices, share findings with other breeders, select stock strategically, plan long-term, and work ethically and cooperatively to preserve and protect several genetic lines of American Guinea Hogs.

Code of Ethics:
HHN members will
- Be truthful in documentation and not sell animals of unknown heritage
- Treat their swine humanely, and in a way to further well-being of the animals
- Not transfer or sell hogs to parties not vetted to look after its health and safety or who may exploit or degrade lines or act in detriment to the Guinea Hog breed
- Only breed or sell animals in good condition and health
- Monitor the herd for potentially adverse effects of known genetically inherited conditions.
- Educate prospective buyers regarding the implications associated with the presence of these conditions in a breeding program
- Work to protect and improve the good standing and reputation of the breed and the Historic Herds Network
- Assist, however possible, in herd dispersals implemented by other network members or when notable herds are brought to their attention.

The three of us raising part of the Sumrall herd decided to add the prefix "MS" in front of names of hogs descended from Sumrall Bobbie Sue. The Sumrall herd was once referred to as the "Mississippi herd." We also gave that prefix to additional generations when they were descended from Sumrall Bobbie Sue. This way, even though Bobbie Sue may not show up on a three-generation pedigree, at a glance someone will see the MS prefixes and identify the pedigree as Sumrall linebred. It is a permanent marker of the hidden genetics. The Sumrall-bred hogs from Dean and Rhonda Shirley's registered stock Gary purchased in 2012 were *not* descendants of Bobbie Sue, and would *not* include the MS marker unless they had one parent or grandparent who were.

Breeders of the HHN are selective about their buyers. After all the effort put into saving the genetics, they do not want them to be on a farm that crossbreeds, to farmers who do not have sturdy field fencing, to someone not zoned agricultural, or to someone who wants a pet. Their livestock is selected to survive on pastureland, actively breed, and to produce quality meat and breeding stock.

Donna pulled together all of her pedigree information and began to put puzzle pieces together in regard to the genetics of her very large herd, as well as helping Deb and I piece together genetics in our herds. Once I sorted through my research, putting together this story, and completing a few more interviews, I had the basic pedigree options for the genetics in Becky's herd, as well. Donna served as our Registrar, so we could put together pedigrees as our sows began to farrow.

Donna now undoubtedly owns the largest and most genetically diverse herd of Genetic Recovery Guinea Hogs in the country, because it includes stock descended from the AGHA founders, Stephen and Hollie Brothers, and the stock Arie obtained from Sumrall, Brown, and Hesters herds. Her original Rudy and Sasha include Baylis blood, so she has every possible foundation strain represented. While other farms may have more breeding hogs, many of Solomon's Wisdom hogs will have their AGHA registration under the Genetic Recovery designation.

DNA Testing

Another invaluable service that Donna provided to the network was to contact the University of California, Davis (UC Davis) laboratory to work with us on DNA testing. Although not required by the AGHA, we determined that we wanted to do this in order to determine relationships between our expanding population of hogs. Outside of the Historic Herds Network (HHN) there seemed to be a lot of confusion about genetic testing for the Guinea Hog. Donna contacted the University of California, Davis (UC Davis) Veterinary Genetics Laboratory to find out what services they could or couldn't provide. Implementing this important step was one more feather in Donna's cap.

UC Davis provides only a "swine parentage and genetic marker" report. It reports on fourteen genetic loci, or SNPs. Then those are analyzed against other DNA samples provided to see which other hogs qualify as a possible dam, sire, or offspring. There seems to be a way to infer a sibling relationship or at least "close relationship" as well, but that is not what they promise to deliver.

In the case of the Maveric herd hogs recovered, we had pedigrees. However, the HHN wanted to confirm those pedigrees to assist us in making educated breeding choices and to provide accurate registration records in the event that our hogs were accepted into the AGHA registry. The Sumrall family and Annette Hesters provided some relationships verbally, but the hogs had not been tagged and there were no written pedigrees. The herd was now divided between 6 states, including some genetics in Mississippi and Florida that were not part of the HHN. The hogs picked up after Gary's death had little to no trail of relationships. The Hesters herd had pedigrees up until 2005, but the hogs remaining in 2015 did not have names, ID, or pedigrees. Our goal was to determine the familial relationships. We wanted to be informed before mating full siblings or sire and daughter, for example.

THE GENETIC RECOVERY PROJECT

To gather DNA samples from our hogs, we used hair follicles. The process was very simple. We called a hog, offered it something to eat, and used a pair of pliers to yank a bunch of hairs from its rump. The hairs would have bits of skin surrounding the hair follicles. These hairs would be sent to UC Davis along with their name, sex, and breed. All of the HHN hogs had been given ear tags for identification with which to match pedigrees and DNA tests.

The test for one hog cost $40 as a one-time fee, but it had to have at least two additional hogs to test against. That made the initial testing for one hog a minimum of $120. If some of those hogs did not have an identified sire or dam, others needed to be tested. Here is where networking came in handy. Each of us tested the hogs in our own herds, whether in Georgia, Indiana, Kentucky, or Wisconsin. When we still lacked some relationships, Donna arranged for testing with boars in Mississippi and Florida in order to rule them out. She absorbed that expense. Thousands of dollars were spent on this part of the process.

Donna had the advantage of receiving pedigrees and tags from the Maveric herd. We had some information from breeders. We used the information we knew to establish a particular order of testing we would follow. We developed a plan—certain hogs would be tested against each other initially. As results came in, additional hogs would be tested against those. It was a time consuming, expensive process. By sharing information, we could test a hog in our herd against hogs in another state without paying additional one-time fees. We had to network to get a full picture of the herds.

As an example, my boar Sumrall MS DuBose qualified as an offspring of Deborah's Sumrall Mississippi Belle. He was tested against four different DNA samples of potential dams. We also tested DNA from three males, but no sire was located. In order to attempt to determine his sire, Donna requested additional samples from the two breeders who chose not to be part of the HHN to see if any of their males may have been his sire. In all, a total of 10 hogs were tested for this one case. DuBose's sire was never identified, meaning that it had been relocated or died prior to leaving the Sumrall farm. While it was frustrating to leave that part of his pedigree blank, it can be surmised that DuBose is all the more valuable to the gene pool because his sire is not represented by any of the other hogs, indicating that he carries unique DNA not already in the gene pool.

By the end of this phase, forty-five DNA samples were collected and paid for by HHN members. Donna paid the largest share of fees because she had the most hogs and tested outside of her herd seeking to complete the pedigrees for each HHN member. She was very committed to the process. Between the Maveric pedigrees, the Shirley pedigrees, pedigrees from Annette, the verbal history, and the DNA testing, most of the hogs had three-generation pedigrees of both dams and sires. DuBose was an exception, with only one half of a three-generation pedigree. This was quite remarkable, and would have been impossible without each network member staying in close communication and working cooperatively for the larger goal of preservation for the national herd.

The Genetic Recovery Process

In November, 2016 the AGHA Board of Directors announced at the annual meeting that the Genetic Recovery *process* was approved by membership. One step of the hurdle was complete. Now for the actual *process and review of our application.*

In the interim, I had bred my sow Sumrall MS Swanee Rose for the second time, to her cousin Sumrall MS DuBose. Their first litter had been born. I was so pleased with this linebred pairing that I repeated it in about six months. I selected shoats to move forward and marketed my unregistered breeding stock as "unregistered with no promise of being registered in the future." I did not lower my prices, but educated potential customers about foundation lines and the inherent value of these hogs. Customers were provided pedigrees from either my software

THE GENETIC RECOVERY PROJECT

or Donna's. I was immensely pleased with the remarkably different conformation, teat placement, calm demeanor, lactation, hardiness, thriftiness, and mothering abilities in my first two litters, and in more that came in subsequent generations. These were exemplary and examples of the Guinea Hog breed.

The road to getting the herds finally accepted was arduous, sometimes contentious, and full of bumps and potholes along the way. Members of the Genetic Recovery committee were understandably feeling the enormity of their decisions. Without the same background knowledge HHN members had amassed, they were at a disadvantage. They were concerned about their perceived risk that these hogs included DNA from feral hogs or hogs of a different breed. If accepted, they could contaminate the herd with different DNA genotypes. Those were valid concerns.

We hoped to address those concerns with the sharing of photographs, documentation of consistent offspring over generations, documentation of history, archival documents, and DNA results. HHN members provided information on multiple generations of hogs to help minimize those concerns. I wanted to show that generation after generation produced hogs with a consistent Guinea Hog phenotype, while DNA genetic markers provided evidence of similar genotypes. I had three generations of Sumrall hogs on the ground.

Becky made these comments about the process. "It's important to have patience, patience, patience. I kept my focus on what I was doing here. I went into the project fully realizing that it [acceptance into the registry] may never happen. If someone else locates a lost herd, they need to do the same thing. Not every hog is going to get accepted, and it may just take a while. You have to go at it for the long haul."

In early November 2017 the waiting for some of us was over. The Sumrall hogs were entered into the registry! Things moved more quickly when the Hesters herd was submitted. That process took only a month. The long wait was over! It would be a few more weeks before registration papers were in our hands, but the HHN members breathed a big sigh of relief.

Table 2
Foundation Hogs of the Genetic Recovery Project **Obtained 2015-2016** **Added to the AGHA Gene Pool in November, 2017** **(Eight Hogs, Six Family Groups)**
1) Billy Frank Brown, breeder in Mississippi Herd Source: Pittsville, Georgia *Foundation Boar*: **Brown's Blue Boy**
2) Dan and Shirley Hale, breeders in Michigan Herd Source: broker in Michigan *Foundation Boar*: **Hesters (Hale) Mork**
3) Marcia and Paul Read, breeders in Pennsylvania Herd Source: Pennsylvania petting zoo and Georgia breeder *Foundation Boar*: **Read Iggy** *Foundation Sow*: **Read Ziggy** Read Iggy and Read Ziggy were probably half-siblings
4) Randy Setty, breeder in Virginia Herd Source: Unknown *Foundation Boar*: **Setty Blackbart** *Foundation Sow*: **Setty Lucky** Setty hogs were closely related
5) Gary Sumrall, breeder in Mississippi Herd Source: Family stock since 1900 in Mississippi and Georgia breeder in 1961 *Foundation Sow*: **Sumrall Bobbie Sue**
6) Unknown Virginia Breeder Potentially J. Frank Baylis *Foundation Sow*: **Hesters Ginger**

Part VIII
Where do we go from Here?

Chapter Thirteen
Keeping our Momentum: A Call to Action

> "Selection is the only tool that can change gene frequencies in a population."
> Sponenberg, Martin, Beranger
> *Managing Breeds*

Know Where you are Going: Developing a Philosophy, Strategies, and Goals for your Herd

As the Historic Herds Network moved into the future, I developed a tool to guide members as we continued our work. *Managing Breeds* offered many examples of breeding strategies, selection, and philosophies. While there are no right or wrong goals or strategies, it is important that you *have* goals and strategies. It is important for your marketing to articulate to your customers what your philosophy is and what goals and strategies you employ.

If a conservation-minded breeder reads this book and decides to linebreed a particular line, that person will want to find a breeder knowledgeable in the line that can provide the stock that person wants. If a person wants to linecross, then two lines must be available to complete that task. If a person just wants to keep a good-natured homestead hog, a wider range of options is available.

In the HHN, we reviewed the various breeding philosophies, strategies, and goals we learned about in our Managing Breeds studies. I wrote out my own model that I shared with customers who purchased my stock from Broad River Pastures. We discussed our varying strategies with each other when we met for our monthly teleconference meetings. As we articulated these among ourselves, we became more confident about our choices.

Breeding Strategies

The purpose of linebreeding and inbreeding is to make the offspring more genetically similar. Any breeding of related animals is inbreeding, although Sponenberg, Martin, and Beranger reserve that term for first-degree relatives. Think about your immediate family. It consists of children, parents, and siblings. Breeding father to daughter, mother to son, and brother to sister is considered inbreeding. If you breed that closely, you can get more consistency in your litters, but you are eliminating many genes when you make your breeding selections. You may also get "the best of the best and the worst of the worst." The good traits will be emphasized, and so will the bad. This is one way to bring the weaker traits to the foreground so you can cull them to remove them from the gene pool. This is how standardized breeds are formed.

Linebreeding can also increase similar genes. It involves the breeding of more distantly related animals. Think of grandparent to grandchild, aunt to nephew, uncle to niece, and cousin to cousin. You will see a uniform look in these offspring because of the family relationship. If you want predictable outcomes and you carefully select the traits you desire, you will have a fairly predictable line, hence homogenous.

The drawback in linebreeding and more so in inbreeding is the risk of a loss in vigor and reproduction. This can sneak up on you without warning. In order to decrease the risk, it is advised to linecross one generation and linebreed the next, still selecting for the more uniform traits, but introducing diversity in alternate generations. Once continual inbreeding for multiple generations affects vitality and ability to reproduce, it is too late to go back. The damage is done.

In an online discussion of the book *Managing Breeds*, D. Philip Sponenberg addressed a question about inbreeding risks. He wrote, "The overall point is that not all inbreeding may be quite the same. Where to draw the line is the question and making sure you can minimize it over most breeds makes sense. BUT, if 'way back' in the pedigree, then it just might not matter as much as mating a brother and sister.

"'Hazardous' does indeed mean inbreeding depression, and that is usually a reduction in general vigor and reproductive traits. That's why it is good to watch inbreeding. My opinion would be that this is especially true of inbreeding done over sequential generations. Inbreeding IS useful for certain situations, just you can't use it as the only strategy!"

He continued, "The general tracking is for growth rates, overall vigor, and reproductive success. To track it, some programs can chart out 'test' matings, and this would be pretty useful in general. That way you can see if outbred matings (those with minimal inbreeding coefficients) are still possible. That can turn into a search, by every breeder, for the 'least related mating,' and that can also have problems down the road because in small breeds, such a strategy works for a few generations, but then everyone is related to everyone. So my overall preference is some combination of inbreeding/linebreeding and outbreeding (linecrossing) from generation to generation. And to make sure that each animal in a breed DOES have an outbred mate, at least potentially.

"Alas, this is going to vary breed to breed, depending on how diverse the founders are. This is why, for so many of our breeds, my constant refrain is to 'find more that are out there.' This doesn't always happen, though."

I suggest again that you think about what goals you have for your herd. Are you breeding to continue a particular line? Do you want your hogs to be known for their excellent temperament? Do you want thrifty piglets that grow on pasture with minimum input? Are you interested in crossing lines for vigor in the first generation? Once you know what your goals are, it will be easier to select your replacement stock or breeders to offer for sale. The important thing is to think ahead and know what your goals are. The animals that do not meet your criteria will taste delicious. For complete guidance in breeding strategies, read *Managing Breeds.*

Selection

We have plenty of numbers in the national Guinea Hog herd now, although most of them remain unregistered. It is important now, more than ever, that selection is deliberate and focused on your goals for your farm, your customers, and the health of the future national herd. Donna Dorminey, Deborah Baker, Becky Mahoney, Gary Sumrall, Kirk Fackrell, J. Frank Baylis, Don Oberdorfer, and I have all reflected on our selection goals and they are recorded in this book. These are helpful examples of breeders beginning with the goal in mind from the start. You can do this, too, if you are a breeder. Create a visual picture and description of what you want.

I created the American Guinea Hog Selection Matrix to familiarize breeders with the range of characteristics acceptable in this landrace breed, and those that are unacceptable, based on the AGHA breed description and interviews with long established breeders. I had been thinking about selection criteria for a while when I read an article written by Alison Martin and Jeannette Beranger called "Buyer Beware!" in *Sustainable Farming* magazine. I followed up that reading with a conversation in which I asked Jeannette how to select appropriately. What did I need to look for specifically in a landrace breed? She referred me to a matrix Phil

Sponenberg had developed for breeders of Colonial Spanish horse breeds. I decided to develop a similar tool for the Guinea Hogs. Several staff members at The Livestock Conservancy plus Phil Sponenberg offered feedback and suggestions during its development. The matrix is a guide to help breeders make selection decisions, and no part of it is prescriptive, due to the breed's landrace status.

When I had my first interview with Don Oberdorfer, I asked him what he wanted breeders today to think about when selecting hogs. He shared the following with me.

"What I think is a real danger for Guineas right now would be to have a standardized look. What we don't want to get into is German Shepherd Dogs with hip problems. So you want different breeders in different situations to breed for different traits. So that someone would say, 'Oh, Cathy Payne's pigs, they are shorter and heavier than others out there,' for example. And 'That guy in California has ones that are a little longer and rectangular and he breeds for that type.' And you want different folks to breed for different things so more of the genes are expressed. So down the road, even if they come from the same ancestors and cousins, there is more distance between them. So if they are fifth cousins there is more distance. I think some danger with the small breed is for them to say 'This is the only way we should breed.' Then we will have a narrow band of characteristics. We want things that are healthy, but in a wider parameter."

He continued, "You want some to be the smaller kind. And you want some that may be the larger kind. I found in the ones that I had originally, there were quite different shapes of them. Being a landrace breed, that is logical. But part of the problem with mixing everything together, you end up with the soup where you have lost the traits of the individual ingredients. What you want breeders to do is start to pull out and look at a wider set of characteristics so we can reassert some of those rare genes. If you institutionalize the breed description, everybody starts going for the same thing. That is where you get into trouble with a small population.

"I'm not saying they need PhDs in genetics, but they do need to be more conscientious about who they're breeding, who they're selling, and what is going back into the gene pool."

Today, in 2019, Don is still at Dodge Nature Center and continues to breed the Guinea Hogs. However, he now breeds "more for meat than for genetics. There are enough other folks breeding for genetics now, so I am happy to leave it to them." Don may not be selecting for genetics himself, but his wise words are recorded as a map to guide breeders now and in the future. Without Don's wisdom, initiative, thoughtful foresight, action, and initiative, we would likely not have the breed today.

There are a few breeders who have used careful selection within a closed herd for several years. I am aware of at least two people since 2005 that kept a closed herd, tracked production and survival traits, culled hard, and thought about selection. One is Arie McFarlen of Maveric Heritage Ranch, who started with diverse genetics but kept a closed herd for nine years. She never introduced a new boar or sow to her initial herd after her acquisitions in 2005. Instead, she selected replacement stock from within her herd. The Maveric bloodline is now registered under the Genetic Recovery project and can be found on Solomon's Wisdom Farm in Kentucky, under Donna Dorminey's watch. Donna has a keen eye for selecting sound stock and is committed to retaining traditional style genetics for future generations. Her herd includes genetics from every bloodline in the AGHA registry. She is concentrating on pulling out the Sumrall component of Maveric breeding.

The second breeder that I am aware of is Kirk Fackrell of Cascade Meadows Farm. To date, he has kept a closed herd for fourteen years. He started his herd with genetics from the herd first owned by Stephen and Hollie Brothers (purchased from Arie) and from the Read and Hale lines of Hesters hogs. His hogs are beautiful representatives of the breed. His genetics fatten on quality pasture and are selected for parasite resistance. He has selected carefully.

I am certain that there are other breeders out there with five or more generations of carefully selected, closed and registered herds of Guinea Hogs with a signature phenotype and particular genotype based on selection criteria. Perhaps there are also pockets of lost unregistered herds hiding in the hills somewhere—the Harper, Cox, and Baylis herds have not been located, for example. Not everyone talks about selection strategies or long-standing closed herds, and not everyone has a social media presence. I'm sure they are out there, busy breeding their hogs. I would love to hear about some modern twenty-first century lines and what traits they have moved forward for the breed. Future breeders will undoubtedly discuss with appreciation the bloodlines in development by today's thoughtful breeders. I certainly hope that will be the case.

Involvement

In order to keep the American Guinea Hog breed part of the American landscape for the next one hundred years, it is vital to have new members joining a well-functioning registry. The AGHA has been tracking the Guinea Hogs for the last thirteen years. Breeders must be members of the AGHA in order to register their piglets. I highly recommend membership. Without documentation, the genetic packages cannot be tracked and managed.

A team of volunteers leads the AGHA. Only the registrar is paid. Consider volunteering or serving on the board, helping to write articles for the newsletter, or serving on a committee. Attend annual meetings held by teleconference and vote for your board members. Check the pulse of the organization. Good people with good people skills are always needed. Rather than criticizing, choose to suggest and implement solutions. If you see something that needs to be improved, get involved. The hogs deserve a well-run organization to keep it going, and the AGHA is only as strong as membership and leadership make it.

If you are not in a position to raise livestock, you can still help to support the breed. Share what you've learned with your friends, family, co-workers, and social networks. Most people are still unfamiliar with Guinea Hogs. Find out if there are any breeders in your area and purchase pork from them. By now you are aware that the culls the breeder deems inappropriate to procreate must be eaten. If inferior stock is allowed to breed, the breed could change in a negative way. In order to save the breed, it must be used for its original intent, which is to make lard and meat. You can also patronize restaurants that have Guinea Hog on the menu. Give your compliments to the chef and thank that person for supporting breeders of this wonderful hog.

I have high hopes for the future of the Guinea Hogs. They are wonderful, thrifty, vigorous, and gentle hogs full of personality. They are easy to love, and easy to handle. They make great companions on the homestead while providing delicious pork and lard. I wrote this book to provide you history, knowledge, and insight based on the wisdom of dozens of former breeders. You've been given a leg up in regard to good stewardship. I hope that many others will pick up the torch where I have left it. Guinea Hogs are an important part of the American culture. With your help, they will continue to not only survive, but to thrive for many decades to come.

About the Author

Cathy R. Payne retired from a thirty-three year teaching career in 2010. At age fifty-seven, she started a sustainable farm, Broad River Pastures, in Elberton, Georgia with her husband, Jon. The farm specialized in nutrient-dense food and heritage breeds of livestock. In her third year of farming, she discovered the Guinea Hog breed. It seemed like a perfect fit for an eleven-acre homestead. She became frustrated, however, about the scant information she could find about the hogs' history and bloodlines.

    Cathy had experience with interviews, research, and writing from her doctoral program, so decided to write a book herself. She has several published newsletter articles written for the American Guinea Hog Association and The Livestock Conservancy. She has been interviewed for several articles on heritage breeds including *The Wall Street Journal* in 2017, *Farmers and Consumers Market Bulletin* in 2016, *Ampersand Magazine* in 2017, a "Herstory" article in *The Ladies Homestead Gathering* E-newsletter, and "Meet an Eco-Farmer," a feature of *Acres USA Magazine*.

Additionally, Cathy has been featured in several prominent podcasts to educate listeners about heritage breeds. She has presented talks about heritage breeds for the South Carolina Organization for Organic Learning, the 40th Annual conference of The Livestock Conservancy, and more. Cathy's farm is featured in a section on heritage breeds in *An Ecosystem Approach to Sustainable Agriculture: Energy Use Efficiency in the American South,* by Carl F. Jordan, Springer Press, 2013.

Cathy is enamored with heritage breeds. She has been an active member of The Livestock Conservancy since 2010 and the American Guinea Hog Association since 2013. This is her first book, and the first in a series about the Guinea Hogs. Cathy recently sold her farm and moved to Athens, Georgia to be near family and community. She now divides her time between writing, consulting, speaking, marketing, and managing her website at www.guineahogbooks.com. Cathy can be contacted by email at guineahogbooks@gmail.com.

Acknowledgements

I was so fortunate to have the late-in-life opportunity of a grand adventure in sustainable agriculture that focused on raising a variety of heritage breeds. While I loved them all, my heart really poured into the Guinea Hogs. I am grateful for the opportunity I had to meet so many wonderful people who shared their stories with me. I have more indebtedness than I could ever repay for those of you who helped me on this journey to record the Guinea Hog history.

To my husband, Jon, who never doubted that I would write this book about the hogs I loved so much. Over the last six years, he helped me out of more technical glitches than I can count!.

To my first supporter, Lisa Naumann, who inadvertently set me down the path of learning the history of this breed. You were at just the right place, at just the right time, with just the information I needed to begin this project.

To Jackie Medina and Pam Allgood, my wonderful friends. You've helped to prop me up during many difficult moments in my journey and rejoiced with me when I met with success. I treasure our friendship.

To breeders who came before me and shared information freely, answering questions for a greenhorn. I took it all in and proceeded to pay it forward to benefit those who came after me.

To members of The Livestock Conservancy who answered my questions, opened up the archives, directed me to sources, and listened to my ideas. You are always supportive and appreciative of my efforts. Jeannette Beranger, Charlene Couch, Alison Martin, and Phil Sponenberg, thank you for your encouragement. In addition, the authors of *Managing Breeds*—Sponenberg, Martin, and Beranger—

were kind enough to give permission for me to quote extensively from their helpful book in this manuscript.

To Chuck Bassett, former executive director of The American Livestock Breeds Conservation. You guided me through some challenging times, offered experienced advice, and supported my work on behalf of the Guinea Hogs.

To every person who answered the phone, spoke to me, and shared your story. It was an honor to record your words, learn from your wisdom, and learn from your knowledge and experience.

To my transcriber Lori Voigt; I couldn't have written this book without the data. It was pure gold to preserve each person's true and accurate voice.

To Matthew Hunker who helped me track down breeders that had eluded me and was always in the corner cheering me on.

To my cohorts in the Historic Herds Network: Deb Baker, Donna Dorminey, and Becky Mahoney. Thank you for sharing this journey with me and for all your continued work on behalf of the Guinea Hogs.

To the interns and farm manager from 2013 to 2017 who helped me care for my precious hogs while I juggled farming and marketing with researching for this book—Dillon Norm, Adam Hegel, Sara Black (she also transcribed the interview with Gary Sumrall), Alex Guerrero, Ashley Dodd, Wesley Dodd, Devin Maloney, Lauren Read, and Bradley Austin (farm manager). We also had notable volunteers in 2018 as we closed out the farm. These included Jerilyn Benefield, Marti Edwards, Miranda Cooper, and Sara Black.

To those breeders who are now caring for my former breeding hogs and breeding offspring that moved on in 2018, I appreciate you more than you could know. Without the peace of mind you provided me, I might have been too heartbroken to continue this work in my new home and preserve this history for the future. I know you will continue the hard but joyous work of maintaining this breed.

ACKNOWLEDGEMENTS

They include Jessica Creighton, keeper of BRP Muddy Waters, BRP MS Autumn Primrose, Sumrall MS DuBose, Sumrall MS Swanee Rose, BRP MS Sister Rosetta, and BRP MS Memphis Minnie. I hope to visit this summer. Sam King, you have a nice herd with lots of Sumrall genetics: BRP MS Baker's Pride, BRP Jackson's Pride, BRP MS Artesia's Pride. May they serve you well. To Mike Bolen who helped BRP Billie Holliday spend half a year at Middleton Place Plantation to educate people about these southern hogs, with thanks to Gra' Moore for transporting Billie to the plantation. Aura Morris and Dave Gladson have my beloved BRP MS Ma Rainey and SWF MS Big Bill Broonzy near me in Athens, Georgia. Thank you for growing out piglets for my freezer and for welcoming my granddaughter and me when we come by to see how everyone is growing.

Janice Agarwal, I hope that BRP MS Ella Fitzgerald and BRP MS Della Reese are growing out well for you in Indiana. Gregory and Lisa Reaves, I'm glad you got my boy BRP MS Beale Street Blues Boy and his girls, BRP MS Ida Cox and BRP MS Mamie Smith. You have a nice mix of genes there. Shannon Englehart, thank you for adopting sweet Yokeley's Summer Thyme and BRP Sonny Boy. Those are two sweet hogs, and photogenic to boot! Sandee House and Bryan Rhodeghiero, I hope that UGF B.B. King and SWF MS Willie Mae do well for you. Nathan Bundrick has some genetics from me, as well, but I can't recall who at this time. I hope I didn't forget anyone else.

The assemblage of talent that has helped me put all the finishing touches on this book has been incredible! Lila Mijailovic designed my book covers and the interior design of my print books to give me a polished look. Gabi Rosenthal's photographs of the hogs and me helped make the cover design pop. Jen Henderson formatted the E-book, taking much stress off my plate. Jennifer Bradshaw was the most amazing editor ever. She understood this project well and was a great collaborator. The print books had the additional help of a professionally produced index by Kelly Burch. I had considerable help from beta readers, proofreaders, Indiegogo funders, and my launch team. My remarkable web master and marketing specialist, Kathryn Baker, helped build the pedigree gallery, pull things together, and coordinate Mail Chimp communication with the launch team.

To the AGHA board of directors and AGHA members who developed and approved the Genetic Recovery policy and those who voted to admit the missing genetics, therefore increasing the biodiversity in the national herd, I appreciate each one that took on the extra responsibility. It will make a huge difference in the hogs for many years to come.

Resources

Organizations and Websites:

American Guinea Hog Association - This organization focuses on preserving American Guinea Hogs and supporting their breeders. Members have access to registration and an informative newsletter. There is an online registry for members only.

    https://guineahogs.org/

Guinea Hog Books and Information - This website provides information about history, breeding, and marketing of Guinea Hogs. It is the author's site for Cathy R. Payne. It includes a pedigree gallery that shows various lines of the hogs as described in this book.

    https://guineahogbooks.com

The Livestock Conservancy - Their mission is "to protect endangered livestock and poultry breeds from extinction." They have monitored and promoted the Guinea Hog for thirty years and are in their forty-second year of operation.

    https://livestockconservancy.org

Bibliography and References:

Sponenberg, D. Phillip, DVM, Jeannette Beranger, Alison Martin. *An Introduction to Heritage Breeds: Saving and Raising Rare-Breed Livestock and Poultry* by The Livestock Conservancy. Storey Publishing, 2014.

Sponenberg, D. Philip and Donald E. Bixby. *Managing Breeds for a Secure Future*: *Strategies for Breeders and Breed Associations*, 1st ed. American Livestock Breeds Conservancy, 2007.

Sponenberg, D. Phillip, Alison Martin, and Jeannette Beranger. *Managing Breeds for a Secure Future: Strategies for Breeders and Breed Associations*, 2nd ed. 5m Publishing, 2017.

Maclay, Kathleen (AP). "Breeding Rare Animals Helps to Preserve Diverse Gene Pools." Los Angeles Times, March 04, 1990.

Wigginton, Elliot, and High School Students, ed. *Foxfire 3*, 1st ed. New York: Anchor Books/Doubleday, 1975.

"African Guinea Hogs Make Great Pets." *Farm Show*, Volume 14, Issue 4, March–April, 1990, 18

Heise, Laurie and Carolyn Christman. *American Minor Breeds Notebook*. The American Minor Breeds Conservancy, 1989.

Improved Essex Swine, Volume III. Published by The American Essex Association, 1896.

Sponenberg, D.P. "Pig Breed Relationships." *The Livestock Conservancy News*, Volume 30, Issue 3, Summer, 2013.

Philpott, Tom. "A Reflection on the Lasting Legacy of 1970s USDA Secretary Earl Butz." *Grist.org* online newsletter, February 8, 2008. https://grist.org/article/the-butz-stops-here/

Estabrook, Barry. *Pig Tales: An Omnivore's Quest for Sustainable Meat*, 1st ed. New York: W.W. Norton and Company, 2015.

Prichard, Diana. "The Rise and Fall of the Great American Hog." *Modern Farmer*, August 21, 2013.

Ferdman, Roberto A. "The Decline of the Family Farm in One Chart." *The Washington Post*, September 16, 2014.

Oberdorfer, Don. "In Praise of the Guinea Hogs." *Small Farmer's Journal*, Volume 28, Number 4, Fall 2004, 60.

Podger, Pamela J. "A Morsel of History: Rare Hog Charms Park Visitors." *Roanoke Times*, May 2, 2005. (reprinted in the Daily Press online)

Ross-Flanigan, Nancy. "Collecting Living Antiques in the Barnyard." *Chicago Tribune*, October 12, 1987.

Martin, Alison and Beranger, Jeannette. "Buyer Beware!" *Sustainable Farming*, Volume 2, Issue 1, Spring, 2017.

Glossary

AGHA: The American Guinea Hog Association founded 2006; includes a registry.

ALBC: American Livestock Breeds Conservancy: Protecting and promoting heritage breed livestock in the United States. Originally the American Minor Breeds Conservancy founded in 1976, and currently The Livestock Conservancy.

American Guinea Hog Association: see AGHA

American Livestock Breeds Conservancy: see ALBC

American Minor Breeds Conservancy, a.k.a. AMBC; see ALBC

bacon type: a long-bodied muscular pig that produces less fatty meat than a lard type

barrow: A male hog castrated before he reaches sexual maturity.

boar: A male pig or hog with testicles intact; generally fertile from four months to twenty years.

chitterlings: The small intestines of a pig, cooked for food; a favorite among southern families, especially during the holiday season.

coefficient of inbreeding/inbreeding coefficient (COI/CI): A measurement in genetics to calculate the relatedness of animals; a higher number indicates a closer relationship.

coefficient of relatedness (COR): The probability that two individuals share an allele (a variant form of gene) due to recent common ancestry.

coronet band a.k.a. coronary band: In a horse, the lighter band circling the top part of the hoof. J. Frank Baylis uses this term as analogous to slim white "socks" on some of his otherwise black Guinea Hogs.

cracklings: After rendering pork fat and straining it, there are brown, crispy bits of meat and skin remaining; these are the edible and delicious cracklings.

crossbreed: A cross between one purebred animal and another as in a Labrador to a poodle.

cull: To eliminate from breeding stock or herd through processing for meat, castration, or sale. The goal is to remove an inferior animal from a herd and maintain healthier, more productive, sound, and thrifty individuals to move genes forward.

dam: The female parent of an animal; often a domestic animal.

feeders: Pigs (typically barrows) sold for someone to grow out at another farm for meat purposes only.

farrow (verb): The act of a sow giving birth to a litter of pigs.

foundation: The original animals in a herd. It can be local (farm level) or national (total registry). Each foundation animal in the herd should leave at least one replacement animal of the same sex in order to maintain diversity in the herd; reducing inbreeding.

genetic bottleneck a.k.a. population bottleneck: An event that causes a dramatic reduction in the population. The cause could be an environmental disaster such as a fire, flood, or global warming. It can also be caused by poor management strategies such as over hunting, destruction of habitat, or multi generational inbreeding.

genotype: A complete set of genes in an organism; unseen make up of cells; expressed outwardly as phenotype (genes determine eye color, but you don't see the genes, you see the eyes).

GHA: Guinea Hog Association (1990s organization).

gilt: A female pig that has never farrowed.

HHN: Historic Herds Network, also the American Guinea Hogs Historic Herds Network. Formed in 2016 by four breeders to keep a registry and share ideas to preserve an unregistered group of purebred

Guinea Hogs. They tested DNA, conducted interviews, and observed multiple generations of hogs until the herds were entered into the AGHA.

hog: Domestic swine, generally one weighing more than 120 pounds. The breed Guinea Hog uses this term at all ages in the breed name to avoid confusing it with the guinea pig, a rodent often raised as a pet or used in scientific research.

inbreeding: Breeding of two animals that are more closely related to others in the breed, such as full siblings or sire to offspring. Such breeding minimizes genetic diversity, a trait of value in landrace breeds. It can result in poor reproduction and poor disease resistance.

lard type: pigs that fatten quickly, typically with short legs, a rotund body, and fatty meat, raised for producing lard; meat is fatty.

linebreeding: Breeding of two animals that are related. A type of inbreeding that might include second cousins, grandmother to grandson, etc. Inbreeding and linebreeding should be followed by a generation of line crossing to infuse genetic diversity and maintain production.

linecrossing: Breeding of two animals who are not related (no common relatives in the first three generations). This strategy can only be used if some linebreeding has also been used. If every breeder has a goal of low COI in a small population, eventually none will have low COI. They will all be related. Linebreeding can be alternated with line crossing.

Minor Breeds Conservancy: An American organization founded in 1977. Its name was later changed to American Livestock Breeds Conservancy and its current name is The Livestock Conservancy.

phenotype: Observable characteristics such as reproduction, color, or temperament, and resistance to disease that result from underlying genotype and interaction with environment.

pig, piggie: Swine. Old-timers use these terms to mean piglet, or young swine, with their dam.

piglet: Young swine.

render: The use of steam, water, or dry heat to melt the fat of a pig and separate the fat from meat. The rendering process yields lard as its final product.

scrapple: Seasoned pork, often including offal, mixed with cornmeal, molded, then sliced and fried.

selection: A vital component of breed management, selection is the choosing of which animals will breed and which will not. In the development of landrace breeds, much selection was environmentally determined, as in "survival of the fittest." On managed farms, it is more manager driven by those raising the livestock. Breeds require planned selection to avoid inbreeding to prevent reproductive bottlenecks. The goal of selection is to produce an adapted population that is both productive for the farmer and viable for the survival of the breed

shoat: A young pig, usually at weaning age, either male or female.

sire: The male parent of an animal; often a domestic animal.

sow: A female swine or hog that has farrowed at least once.

thrifty: Growing vigorously.

The Livestock Conservancy: see ALBC and Minor Breeds Conservancy.

UC Davis: University of California, Davis Campus. Base for the Veterinary Genetics Laboratory that tested DNA to determine relationship status between more than forty hogs evaluated by the Historic Herds Network and submitted to AGHA Genetic Recovery project to increase genetic diversity in the national herd of American Guinea Hogs

Appendix

Appendix Documents:

GHA Breed Description	195
AGHA Breed Description Early Draft	197
AGHA Breed Description	199
AGHA Breeder Prefix Guide	201
Foundation hogs in the AGHA	202

Guinea Hog Association's Guinea Hog Breed Definition/Standard published Winter, 1991 in the GHA Newsletter
Editors, Lola Moffit and Gabriella Nanci

Traits that the ideal Guinea Hog should possess:

Back - straight, not swayed
Tail - curled or bent
Legs and Feet - should easily support body weight
Toes - not splayed
Skin - smooth and tight, not gathering in folds or wrinkles
Hair - rather long and plentiful, covering entire animal
Ears - large, pointed, and erect
Eyes - bright and prominent, not covered by folds of fat or skin
Color - solid black
Weight - between 60-120 pounds
The ideal Guinea Hog will have a **pleasant disposition**.

Traits that are acceptable but not ideal:
Hair - sparse, uneven, or very coarse
Feet and toes that show some signs of 'splaying'
Weight - over 120 pounds, or under 60*
Color - small white markings, limited to feet or face

Traits that are unacceptable:
Back - swayed
Feet and Legs - "breaking down;" cannot support the pig's body weight (also known as weak hocks)
Tail - straight with no bend
Snout - "pushed in" or excessively short
Color - anything other that solid black or black with a few white markings
Skin - wrinkled skin or skin that hangs in folds
Weight - over 200 pounds or under 40 pounds
Ears - small round ears or ears that flop over
Temperament - any signs of aggression

* Keep in mind that pigs are famous for weight gain, and will often fluctuate between categories. We have attempted to set a standard that discourages both over feeding and under feeding, as well as breeding for extreme sizes. Pigs can be of types A, B, or C, with all types being of equal merit.

Type A has a narrow head and a light bone structure.

Type B is essentially the medium of Types A and C, but is usually distinct with its broad forehead tapering toward the muzzle and a medium dished nose.

Type C has a broad head, a short, broad nose, and a heavier bone structure.

(Pigs can be of breed types A, B, or C, all types being of equal merit)

American Guinea Hog Breed Description
Draft Copy by Kirk Fackrell, 2006

The American Guinea Hog is the ideal sustainable heritage farm pig. It is known for its perfectly moderate size, excellent foraging abilities, exceptionally friendly temperament, excellent meat and indispensably flavorful lard for cooking. While the American Guinea Hog is smaller than industrial hog breeds, it is a good-sized farm pig providing a nice, very well marbled carcass.

Origin: The American Guinea Hog is a true American heritage breed of domestic farm pig, perhaps over 200 years old, with debatable distant European, African, and perhaps even Chinese ancestral influences.

Height: Adult American Guinea Hogs (at 2 years of age) range from 24 to 30 inches tall at the back, adult males averaging 1-2 inches taller than females. Older animals may grow larger.

Body/Length: Fully grown adult American Guinea Hogs range from 46 to 58 inches, measured from a point between the ears to the base of the tail. They have a straight back with little or no arching. From a side view, they should present a long, rectangular appearance (with rounded corners).

Weight: Well-conditioned, fully adult American Guinea Hogs range from somewhat under 200 pounds to somewhat over 300 pounds, depending on sex, frame-size, and body condition. Young feeder pigs may reach 120-150 pounds (60-75 pound carcass) at six months under optimal conditions. Because American Guinea Hogs easily fatten, care should be taken to NOT overfeed, especially with grain. Excess weight could lead to fertility problems and feet problems.

Head/Face/Ears/Tail: American Guinea Hogs have medium-small sized, upright ears (sometimes slight bent at the tips in adults). They have slightly dished faces with snouts that vary from rather short to medium-long. The tail has a single curl.

Color: Most American Guinea Hogs are solid black. A common variation (due to a widely spread recessive gene) is solid black with minimal white points at the feet and tip of nose. Excess white (beyond the feet and the end of the snout) is discouraged. An extremely rare recessive red gene exists in the breed, and may rarely exhibit.

Hair: Adult American Guinea Hogs typically have medium to long, coarse, bristled, black hair, some tinged in reddish-brown tones or bluish-black tones.

Temperament: The American Guinea Hog is exceptionally calm and friendly making it an excellent choice for small sustainable family farms. They have exceptional mothering skills. Males and females with piglets are easily managed. They do well with children and a wide range of farm animals.

Life Expectancy: The expected life span of the American Guinea Hog is 10-15 years.

Living Environment: While American Guinea Hogs are suited to wide variety of environments, they prefer lush pastures with clover along with access to minerals, kitchen scraps, quality hay in winter, clean water to drink, access to a muddy wallow, minimal shelter from precipitation and wind, dry bedding, and perhaps a small amount of commercial hog feed or grain. They thrive where ranging and grazing is a constant activity giving them plenty of exercise. They are minimal rooters when good grazing and adequate feed is available.

American Guinea Hog Breed Description
Official American Guinea Hog Association Description from January 25, 2009 to present (2019)

The American Guinea Hog is the ideal sustainable heritage farm pig, known for its moderate size, excellent foraging abilities, friendly temperament, excellently flavored meat and indispensable lard. While the American Guinea Hog is smaller than industrial hog breeds, it is a good-sized farm pig providing a nice, well-marbled carcass.

Origin: The American Guinea Hog is a true American heritage breed of domestic farm pig, perhaps over 200 years old. They developed as a landrace breed throughout the southeastern states of the USA. Anecdotal evidence suggests a European ancestry with other possible influences. It has been determined though genetic testing that the American Guinea Hog is a distinct breed.

Height: Adult American Guinea Hogs (at 2 years of age) range from 22 to 27 inches tall, adult males sometimes averaging one or two inches taller than females. Older animals may grow larger.

Body/Length: Fully grown adult American Guinea Hogs range from 46 to 56 inches, measured from a point between the ears to the base of the tail. They have a straight to slightly arched back. From a side view, they should present a long, rectangular appearance (with flat sides and rounded corners). As a landrace breed, variations are common. Some hogs will be taller and broader at the shoulders with slightly lower and narrower hips.

Weight: Well-conditioned, fully adult American Guinea Hogs range from 150 pounds to 300 pounds, depending on sex, frame-size, and body condition. Because American Guinea Hogs easily fatten, care should be taken to NOT overfeed, especially with grain. Excess weight will likely lead to fertility problems.

Head/Face/Ears/Tail: American Guinea Hogs have medium-small sized, upright ears (sometimes slight bent at the tips in adults). They have slightly dished faces with snouts that vary from rather short to medium-long. The tail has a single curl. Slightly forward facing eyes are common.

Color: Most American Guinea Hogs are solid black. A common variation due to a widely spread recessive gene, is solid black with minimal white points at the feet and tip of nose. Excess white (beyond the feet and the end of the snout) is discouraged. An extremely rare recessive red gene exists in the breed, and may rarely exhibit.

Hair: Adult American Guinea Hogs typically have medium to long, coarse, bristled, black hair, some tinged in reddish-brown tones or bluish-black tones.

Temperament: The American Guinea Hog is exceptionally calm and friendly making it an excellent choice for small sustainable family farms. They have exceptional mothering skills. Females with piglets are easily managed, as are adult males. They do well with children and a wide range of farm animals. It should be a goal of breeders to maintain the good temperament of the American Guinea Hog.

Living Environment: While American Guinea Hogs are suited to a wide variety of environments and will do better than most breeds on low grade forage, they prefer lush pastures with clover along with access to minerals, kitchen scraps, quality hay in winter, clean water to drink, access to a muddy wallow, minimal shelter from precipitation and wind, dry bedding, and perhaps a small amount of grain. They thrive where ranging and grazing is a constant activity giving them plenty of exercise. They are minimal rooters when good grazing and adequate feed is available.

Life Expectancy: The expected life span of the American Guinea Hog is 10-15 years or until they are ready for culling and slaughter as farm livestock, providing excellent meat for the table.

Carcass: At six months, the American Guinea Hog may provide a nicely marbled carcass of up to 75 pounds hanging weight of gourmet-quality highly-flavored meat. Fat tends to peel easily from the meat. There is no need to castrate young male hogs intended for slaughter at six months of age. 1/25/2009

APPENDIX

Guide to AGHA Breeder Prefix

This list includes most breeding prefixes in pedigrees of the first 400 documented Registered and non registered hogs recognized by the AGHA between 2002 and 2009, as well as the 4 breeders in the Historic Herds Network from 2016 to 2017. HHN Members are marked with *.

Prefix	Farm Name	Breeder Name
Audacious		Ethan Schulze
Back 40	Back 40 Farm	Miranda Yearington
Baylis	Bayshore Kennel	J. Frank Baylis
Beardsley	Beardsley Zoo	staff
Biggers	unknown	Bill Biggers
Black Oaks	unknown	James and Anna Perkins
Brambleberry	unknown	1Espri Beauregard
Brown's	Cowpen Creek	Billy Frank Brown
BRP	Broad River Pastures	Cathy Payne*
Brothers	Brothers Farm	Stephen and Hollie Brothers
Cascade	Cascade Meadows	Kirk Fackrell
CHF	Carolina Heritage Farm	Graeme Moore (Gra')
Celesky	Boarding House Farms	Mark Celesky
Chapel Top	Chapel Top Heritage Hope Farm	Deborah Baker*
Crows Nest	Crows Nest Farm	Carmen Smith
DNC	Dodge Nature Center	Don Oberdorfer
Fall	unknown	Kevin Fall
FBF	Foggy Bottoms Farm	John Ross Jr.
Flint and Steel	F & S Farm	Angela Ingraham
GRF	unknown	Tracy Ginder
HeartsEase	unknown	Frances Smith
Hesters	Hesters Farm Log Homes	Annette Hesters
JNH	Home-N-Stead	Becky Mahoney*
LSF	Laughing Stock Farm	Lisa Duff
Mahsrow	unknown	Valerie Worsham
Maveric	Maveric Heritage Ranch	Arie McFarlen
OGS	unknown	Gary Bates
Old Freedom	Old Freedom Church	unknown
Pleasant Ridge	unknown	Brian Yelverton
Rudugast	Rudugast Farm	Gary Bates
SCZ	Sedgewick County Zoo	Callene Rapp
Setty	unknown	Randy Setty
Sugars	unknown	Jessica Benson
Sumrall	Sumrall Farm	Gary Sumrall
Sunrise	Sunrise Farm	Siegfried Forster
Skyfire	Skyfire Garden Seeds	Paul Krumm
Sullbar	Sullbar Farm	Shirley Sullivan & Jim Barnett
SWF	Solomon's Wisdom Farm	Donna Dorminey*
VAZ	Virginia Zoo	staff

201

Table 1
Foundation Hogs of the Early AGHA Registry **Seeded the Gene Pool from 2002-2009** (Twelve Hogs, Three Family Groups)
1) <u>J. Frank Baylis</u>, breeder in Virginia Herd Source: Alabama breeder *Foundation Boar*: **Baylis VA Samson**
2) <u>Bill Biggers</u>, breeder in Virginia Herd Source: unverified *Foundation Boar*: **Biggers Arthur**
3) <u>Randy Setty</u>, breeder in Ohio and <u>Mark Celesky</u>, breeder in Nebraska Herd Source: unknown Mark purchased hogs from Randy and possibly one from Donna Watkins in Illinois *Foundation Boars*: **Celesky's Boris** **Setty Houdini**** **Setty's MC Little Old Stiff Guy*** **Setty's MC Big Old Stiff Boar*** *Foundation Sows*: **Celesky's Lola** **Celesky's Roxanne** **Celesky's Tulip** **Setty Lilly** **Setty Rose**** **Setty's MC Wart Side Sow*** *Littermates born 4/15/2002 **Full siblings

APPENDIX

Table 2

**Foundation Hogs of the Genetic Recovery Project
Obtained 2015-2016
Added to the AGHA Gene Pool in November, 2017
(Eight Hogs, Six Family Groups)**

1) <u>Billy Frank Brown</u>, breeder in Mississippi
   Herd Source: Pittsville, Georgia
   Foundation Boar: **Brown's Blue Boy**

2) <u>Dan and Shirley Hale</u>, breeders in Michigan
   Herd Source: broker in Michigan
   Foundation Boar: **Hesters (Hale) Mork**

3) <u>Marcia and Paul Read</u>, breeders in Pennsylvania
   Herd Source: Pennsylvania petting zoo and Georgia breeder
   Foundation Boar: **Read Iggy**
   Foundation Sow: **Read Ziggy**
   Read Iggy and Read Ziggy were probably half-siblings

4) <u>Randy Setty</u>, breeder in Virginia
   Herd Source: Unknown
   Foundation Boar: **Setty Blackbart**
   Foundation Sow: **Setty Lucky**
   Setty hogs were closely related

5) <u>Gary Sumrall</u>, breeder in Mississippi
   Herd Source: Family stock since 1900 in Mississippi and Georgia breeder in 1961
   Foundation Sow: **Sumrall Bobbie Sue**

6) <u>Unknown Virginia Breeder</u>
   Potentially J. Frank Baylis
   Foundation Sow: **Hesters Ginger**

Index

A

Acorn Eater, 16
Addlestone, Carole, 18–19
Africa as possible origin of Guinea Hogs, 17, 22, 23, 197
African Guinea Hog, 16, 59, 73, 113, 114
African Miniature, 16, 61
African Pygmy, 16
AGH (American Guinea Hogs). *see* Guinea Hogs, 17, 18, 23, 139, 141, 147, 150
AGHA (American Guinea Hog Association), 119, 133, 160, 162, 168, **193**
   American Guinea Hog Breed Description, 199
   Articles of Incorporation, 65
   Board of Directors, 65, 150, 170
   breed standard, 25
   breeders, 79, 139, 201t
   bylaws, 15
   definition, **191**
   formation, 14, 17, 64, 93, 97, 98
   foundation hogs of, 66, 67t, 78, 84, 167, 172t, 202t, 203t
   Genetic Recovery project, 170, **194**
   guidelines, 65
   led by volunteers, 181
   members, 29, 30, 35, 119, 127, 150
   organizing, 62–66
   precursor to, 135
   recordkeeping, 83
   registry of Guinea Hogs, 4, 26, 27, 28, 65, 66, 105, 153
   standards used by, 61, 87, 142
   website, 126, 189
   work to return missing genetics, 1
AGHA Genetic Recovery program, 27, 108, 143, 167
   foundation hogs of, 172t, 203t
   and genetic relationships, **194**
   Maveric line registered under, 180
Agricultural and Food Statistics Report, 54
agricultural systems, 2, 13
agricultural zoning, 167
Alabama, 67t, 78, 79, 81, 129, 202t
Alaska, 11
ALBC (American Livestock Breeds Conservancy), 19, 20, 64. *see also* AMBC (American Minor Breeds Conservancy); Livestock Conservancy, The
   definition, **191**, **193**
Albert, Tammy, 18
Allgood, Pam, 128
AMBC (American Minor Breeds Conservancy), 16, 21, 22, 60, 61, 104, 111, **191**. *see also* ALBC (American Livestock Breeds Conservancy); Livestock Conservancy, The
American Blue and White meat rabbits, 150
American Guinea Hog Selection Matrix, 150, 178
American Guinea Hog Stewardship, 62
American Guinea Hog Yahoo Group, 127
American Guinea Hogs Historic Herds Network. *see* HHN (Historic Herds Network) 192
American Minor Breed Conservancy, 110
Andrew, 100, 101
Angus cows, 116
Archer, Cohen, 7, 8–10, 47
Ark of Taste (Slow Foods), 12, 18
Audacious (prefix), 201t

B

back (body part), 90
   in breed definition, 195
   in breed description, 196, 197, 199
Back 40 Farm, 201t
Back 40 (prefix), 201t
bacon, 21, 41, 42
bacon type, **191**
Baker, Deborah, 137, 138–143, 178, 186
   as breeder, 201t
   recovering the Guinea Hog, 152, 160, 164, 167, 169
   selection strategies, 138, 142, 145
Barnett, Jim, 35, 59, 65, 76, 78, 88
   as breeder, 201t

barrow, 41, 66, 153, 157, **192**
  definition, **191**
Bates, Gary, 201t
Baylis (line), 25, 75, 78, 135, 141, 144, 167, 181, 201t
Baylis (prefix), 201t
Baylis, J. Frank, 32, 67t, 75, 76, 79–81, 107, 109, 172t, 178, **191**, 201t, 202t, 203t
  selection strategies, 80–81
Baylis VA Samson, 26–27, 32, 67t, 75, 75–79, 76, 78, 202t
Bayshore Kennel, 79, 109, 201t
Beardsley (prefix), 201t
Beardsley Park Zoo, 26, 27, 36, 73, 92, 201t
Beauregard, Espri, 201t
belt (coloring), 29, 35
Benson, Jessica, 87, 201t
Beranger, Jeannette, 13, 19, 25, 125, 131, 132, 133, 159, 175, 176, 178
Berkshire (breed), 23, 33, 34, 80
Betty, 104
Big Momma (Sumrall MS), 131, 141, 152
big-boned variety, 8, 23, 41, 74, 88, 119, 120. *see also* little-boned variety
  weight of, 15, 59–60, 118
Biggers (line), 97, 98, 135, 147, 201t
Biggers (prefix), 201t
Biggers, Bill, 62, 63, 67t, 71–74, 201t, 202t
Biggers Arthur, 63, 64, 67t, 71, 72, 73, 74, 75, 78, 87, 89, 202t
biodiversity, 2, 125, 134, 152, 162
Bixby, Donald E., 22, 61, 64, 128, 134
Black Canary Island (breed), 24
black hogs, 12, 26, 30, 56, 61, 103, 119, 122
Black Oaks (prefix), 201t
Black Suffolk (breed), 23
bloodlines, 79, 104, 117, 118, 137, 163, 181
  foundational, 67t, 108, 172t, 202t, 203t
  of Guinea Hogs, 11, 15
  Hesters, 15
  hitting genetic bottleneck, 98
  Maveric, 180
  old Alabama, 81
  of old stock, 140
  preserving, 98, 135
  pure, 48
  Read, 110, 133
  red hogs, 26
  Setty, 15, 141
  Sumrall, 151, 153
  unregistered, 162
  white in, 35
blue hogs, 26, 27, 28, 29, 36, 44, 92, 100, 101, 103, 104, 119, 122
Bluefaced Leicester sheep, 79, 81
Boarding House Farms, 83, 201t
Boarella, 80, 81
boars, 27, 32, 132, 141
  definition, **191**
  personality of, 74, 92
  size of, 15
  with tusks, 71, 73
bone structure, 15, 61, 145. *see also* big-boned variety; little-boned variety
  in breed description, 196
Brambleberry (prefix), 201t
breed, small, 54, 86, 177, 179
breed associations, 1, 4, 98, 127
breed classes, 13, 14, 159
breed definition, 15, 30, 56
  of AGHA, 195–196
breed description, 14, 15, 16, 25, 26, 28, 30, 31, 32, 34, 59, 61, 178, 199
  of AGHA, 197–198
  institutionalization of, 179
breed selection, 33, 178–181. *see also* American Guinea Hog Selection Matrix; selection
  all breeds need, 30
  can change gene frequencies, 175, 176
  characteristics for, 34, 36, 148, 150, 164
  definition, **194**
  difficulty making, 29
  drives for, 13, 25, 28
  and inbreeding, 32, 93
  process of, 156
breed standard, 14, 27, 80, 157
  of AGHA, 25, 61, 87, 142
  of GHA, 25, 32, 61, 195–196, 199–200
  for landrace breeds, 15
breed types A, B, C, 61, 196
breeding pairs, 72, 138, 144, 146, 147, 148, 154
breeding philosophies, 36, 175–176
breeding prefixes, 201t
breeding programs, 4, 94, 166
breeding records, 62, 84, 106, 134, 140.

see also pedigrees
breeding stock, 104, 109, 136, 137, 143, 167
  boars, 80, 85
  and culling, **192**
  eliminating aggressive hogs as, 148
  Hesters line of, 149
  price of weaned shoats for, 60
  selecting, 14, 25, 28–36, 80–81, 93, 106, 108, 115, 117, 130, 134, 138, 142, 146, 148–149
  sows, 8, 80, 85
  Sumrall line of, 155
  unregistered, 170
breeding strategies, 4, 36, 94, 135, 136, 141–142, 149, 152, 166, 175, 176–178
  linecrossing, **193**
  in *Managing Breeds*, 160
breeds, endangered, 18, 22, 133, 138, 146, 189
breeds, rare, 2, 3, 19, 22, 30, 54, 60, 66, 114
  protecting, 79
  pure Guinea Hogs are, 115
Brinson, Freddie, 44
Broad River Pastures, 150, 160, 176, 201t
Brock, Sean, 19
Brothers (prefix), 201t
Brothers, Hollie, 63, 98, 107, 167, 180, 201t
Brothers, Stephen, 63, 98, 107, 167, 180, 201t
Brothers Brewster NR025, 107
Brothers Bronson NR28, 107
Brothers Chunky 2 NR013, 108
Brothers Farm, 64, 107, 147, 201t
Brothers Patti (Pearl) NR004, 108
Brothers Penny NR005, 108
Brown, Billy Frank, 27, 39, 41, 98–103, 162, 172t, 201t, 203t
Brown, Jess, 98
Brown's (line), 162, 167
Brown's (prefix), 201t
Brown's Blue Boy, 27, 103, 162, 172t, 203t
BRP (prefix), 201t
BRP MS Artesia's Pride, 103
BRP MS Memphis Minnie, 103
BRP MS Willie Dixon, 103
BRP Sonny Boy Williamson, 74, 178
Bundrick, Don, 39, 42, 47, 48–49
Bundrick, Nathan, 103
butchering, 41–44, 105, 138
  by corporations, 53
  as family affair, 1
  hogs' readiness for, 200
  reasons for, 31, 32, 35, 80, 139
  weight of hog at, 120
Butz, Earl, 50, 51

C

CAFO (Confined Animal Feed Operation), 52, 53, 54
California, 60, 63, 90
cannon bone, 145, 155
Carolina Heritage Farm, 19, 153, 201t
Cascade (prefix), 201t
Cascade Meadows Farm, 17, 32, 34, 135, 180, 201t
castration, 31, 41, 65, 107, **191**, **192**, 200
cattle, 14, 17, 39, 100, 115
Celesky (line), 74, 97, 108, 135, 139, 144, 147
  have a trace of Watkins line, 109
  hogs of, 27
  linecrossing with Setty, 94
  Mark Celesky trading vs. breeding, 84–85
  relatedness to Setty, 84–86
  at Sedgwick County Zoo, 92–93
  white in, 26, 87
Celesky (prefix), 25, 201t
Celesky, Mark, 62, 63, 67t, 83–85, 90, 97, 109, 201t, 202t
Celesky's Boris, 67t, 92, 202t
Celesky's Lola, 66, 67t, 92, 202t
Celesky's Roxanne, 67t, 92, 202t
Celesky's Tulip, 63, 64, 67t, 85, 92, 202t
Chapel Top Heritage Hope Farm, 138, 153, 201t
Chapel Top (prefix), 201t
charcuterie, 19
Charleston Slow Food Chapter, 18, 19, 20, 24
CHF (prefix), 201t
chickens, 80, 138, 144
  heritage breeds, 146, 150
Chinese origin breeds, 12, 24
  in breed description, 197
chitterlings, 12, **191**
Choctaw (breed), 22, 53
Christman, Carolyn, 60, 61
closed herd, 79, 133, 135, 158, 159, 181
  Baylis, 32, 80
  Dorminey, 163
  Hale, 104

Hesters, 107, 147
Maveric, 180
Read, 132
Sumrall, 163
coat, 10, 60, 80, 92, 93, 140, 141. *see also* hair
   coloring, 27, 29
   coloring variations, 25–27
   types, 74
COI/CI (coefficient of inbreeding/inbreeding coefficient), 93, 94, 98, 135, 140, 163, 177, **191**, **193**
Collins, Fran, 91
Colonial Spanish horse breeds, 179
color, 26, 27–30, 31, 34, 36, 80, 115, 142
   in breed description, 32
   controversies about, 35
   eye, 10, 27, **192**
   fads, 33
   genetics, 28
   Guinea Hog breed standard, 195, 196, 198, 200
   in Guinea Hogs, 16, 25–27, 30
   historical, 23
   meat, 18, 20, 21
   and phenotype, **193**
   traits, 28, 29
Colorado, 64
commercial breeds, 11, 14, 149
commercial production, 2
communication networks, 65, 133, 153, 160, 162, 163, 170. *see also* HHN (Historic Herds Network)
Connecticut, 26, 73
conservation, 2, 59, 139, 146
   genetic, 62
   of Guinea Hogs, 24, 65, 105, 122, 166, 175
   of landrace breeds, 134, 159
   practices, 166
   of wildlife, 150
conservationists, 14
conservators, 54, 63, 106, 109, 122, 166
consumers, 4, 18, 47, 50–54, 52
COR (coefficient of relatedness), **191**
coronary band. *see* coronet band
coronet band, 32, 80, 81, **191**
Cowpen Creek Farm, 41, 98, 99, 103, 201t
Cox, Jean, 63, 181

cracklings, 21, 41, 43
   definition, **192**
Crandall, Sam, 138
Creighton, Jessica, 103
Criollo stock, 17, 98
crossbreeding, 28, 29, 30, 56, 160
   definition, **192**
Crows Nest Farm, 201t
Crows Nest (prefix), 201t
C.S. Foundation Farm, 3
C.S. Fund, 60
culling, 30, 80, 93, 142, 146, 147, 148, 155, 156, 157, 164, 176, 180, 182, 200
   definition, **192**
curing (meat), 11, 43, 44
Cypress Restaurant, 19

D
dam, 75, 87, 94, 107, 160, 168, 169, 170, **194**
   definition, **192**
deformities, 93, 141, 161. *see also* weaknesses
Deihl, Craig, 19, 20
Delilah, 76
deworming, 34, 140
Dexter cattle, 104, 110, 114, 115, 121
disease, 53, 54
   hog cholera, 112
   resistance to, 14, 22, **193**
disposition. *see* personality
DNA, 34, 78, 93, 94, 133, 157, 171, **193**, **194**
   gathering samples, 169
   of Guinea Hogs, 3, 21–24
   testing, 31, 168–170
   verification, 29
DNC (Dodge Nature Center), 62, 63, 64, 74, 84, 87, 89, 107, 139, 180, 201t
DNC (prefix), 201t
DNC Chunky, 64
DNC Esmerelda, 64, 107, 148
DNC Gabriella, 64
DNC George, 63, 64, 107
DNC Junior, 63, 74
dogs, 13, 14, 80
Doherty, Maureen, 113
Dorminey, Donna, 103, 137, 143–145, 178, 180, 186
   as breeder, 201t

recovering the Guinea Hog, 152–157, 160–165, 167–171
selection strategies, 145, 156
Duff, Lisa, 201t
Duroc (breed), 24

E
ears, 114, 145
in breed description, 195, 196, 197, 199
color, 156
for consumption, 43
erect, 10, 23, 90, 91, 147, 148
forward facing, 91, 148
pointed, 60
shagginess, 155
size, 141
tagging, 40, 101, 157, 169
endangered breeds, 18, 22, 133, 138, 146, 189
Engelhardt, Shannon, 74
English breeds, 22, 24
Eurasian wild hog (breed), 29
European origin breeds, 12, 14, 23, 24, 199
in breed description, 197
extinction, 1, 2, 3, 23, 65, 134
of blue hogs, 122
of Improved Essex line, 24
Livestock Conservancy's mission to prevent, 189
eyes, 147
blue, 10, 26, 27
in breed description, 195, 200
brown, 10, 77
coloring, 10, 26, 27, 141, **192**
position of, 10, 89

F
F & S Farm, 201t
face, 80, 141, 148
in breed description, 196, 197, 199
dished, 23, 197, 199
Fackrell, Kirk, 17, 32–34, 135, 178, 180
in breed description, 197
as breeder, 201t
selection strategies, 33–34
factory farms, 50, 54
fads, 27, 28, 33, 55, 115
fainting goats, 79, 142
Fall (prefix), 201t
Fall, David, 201t

Fall, Kevin, 59, 76, 83, 84, 85, 86, 87, 88, 90, 91, 92, 109, 117, 133, 201t
helps form AGHA, 30, 64, 65
selection strategies, 31
family farms, 112, 127, 141, 198
decline of, 49, 50, 54–55
Farm Census, 2012, 54, 55
farm pig, 197, 199
farming, industrial, 47, 48, 49
farmsteads, 3, 56
farm-to-table restaurants, 3, 11, 129, 148
farrowing, 90, 122, 146, 151, 156, 160, 165, 167. *see also* piglets; sows
death at, 76, 88
definition, **192**
in definition of sow, **194**
unassisted, 141
fats, 17, 20, 200. *see also* lard
consumer demand for, 50
in cracklings, **192**
of Guinea Hog meat, 18, 51–54, 105
pork needs for flavor, 49
push for diet low in, 56
rendering, **194**
FBF (prefix), 91, 94, 201t
FBF Sycamore Bonnie, 91, 94
FBF Sycamore Clyde, 91, 94
feed, 86, 105, 139, 200
antibiotics added to, 54
commercial, 11, 198
contaminated, 156
cost of, 157
shell corn as, 121
supplemental, 34
feeders, 43, 131, 138
definition, **192**
size, 197
Sumrall, 157–158
feet (body part), 25, 43
in breed description, 195, 196, 197, 198, 200
and stance, 145
white, 25, 26, 29, 30, 31, 32, 33, 34, 35, 80, 87, 111
feral breeds, 13, 74, 80, 100, 171
Fernandez, Marcos, 18, 157, 158
fertility, 89, 90, 94, 98, **191**
in breed description, 197, 199
fetlock, 155
Fjord horse, 63
Flint and Steel (prefix), 201t

floppiness, 9, 91, 142, 148
Florida, 18, 40, 157, 158, 168, 169
Florida Cracker, 17
Foggy Bottoms Farm, 86, 201t
foraging, 11, 12, 17, 22, 53, 87, 105
   in breed description, 197, 199, 200
Forest Guinea Hog, 16, 17, 22, 71, 73
formal registries, 4, 61, 109
Forster, Siegfried, 201t
foundation hogs, 74, 85, 88, 90, 92, 93, 132, 133, 138, 167, 170
   of the AGHA, 66, 67t, 78, 84, 202t
   boars, 32, 67t, 108, 172t, 202t, 203t
   definition, 26, **192**
   diversity of, 177
   and genetic diversity, 135
   of the Genetic Recovery Project, 172t, 203t
   sows, 67t, 172t, 202t, 203t
founders. *see* foundation hogs
France, 99
Fred, 104, 105
French, Tim, 72

G
Garfrerick, David, 129
Gear, Robert, 16
gene pool, 1, 63, 64, 67t, 93, 98, 117, 139
   of AGHA, 172t, 202t, 203t
   broad, 162
   keeping narrow, 97
   limited, 147
   removing weak traits from, 176
   unique DNA in, 170
gene variants, 27, 34, **191**
genes, 28, 34
   of Celesky's Lola, 92
   conservation of, 2, 13
   and culling, **192**
   deleterious, 151
   dominant, 26, 79
   eliminating, 176
   expression, 179
   frequencies, 175, 180
   and linebreeding, 176
   lost, 54
   rare, 26, 179, 198, 200
   recessive, 26, 29, 35, 198, 200
   weak, 160
genetic abnormalities, 4
genetic analysis, 23

genetic bottleneck, 98, **194**
   definition, **192**
genetic changes, 116
genetic discussion group, 137
genetic diversity, 2, 4–5, 54, 65, 125, 135, 149, 167, 180, **193**, **194**
genetic field, 152
genetic influence, 66
genetic information, 2, 3, 149
   loss of, 125
genetic lines, 78, 127, 128, 136, 141, 142, 166
   rare, 49
genetic management, 138
genetic markers, 168, 171
genetic material, 66
genetic modification, 18
genetic packages, 2, 13, 181
genetic potential, 65
genetic preservation, 54, 62, 98
genetic recovery process, 170–171
Genetic Recovery Project, 27, 103, 143, 153, 170–171, 180, **194**
   foundation hogs of, 172t, 203t
genetic relationships, 22, 85
genetic reservoir, 22
genetic resources, 10, 14, 134, 159
genetic similarity, 143, 176
genetic strains, 78, 133
genetic strength, 24, 25, 97
genetic testing, 83, 168, 199
genetic traits, 13, 28
genetic variability, 22
genetically inherited conditions, 166
genetics, 13, 79, 103, 148
   and breeding, 74, 90, 91, 180
   coefficient of inbreeding in, **191**
   of color, 28
   of Guinea Hogs, 16, 22, 23
   hidden, 167
   improvement in, 151
   preservation of, 122, 167, 180
   tracking, 143
genotype, 13, 24, 27, 171, 181, **193**
   definition, **192**
Georgia, 7, 18, 27, 28, 39, 49, 62, 100, 103, 112, 149
   blue Guinea Hogs in, 101, 102
   farming changes in, 47
   as herd source, 172t, 203t
   stock from, 119, 120, 146

INDEX

GHA (Guinea Hog Association), 25
   definition, **192**
   formation of, 16, 164
   recordation for, 61, 62
gilt, 17, 26, 104, 107, 128, 131, 132, 139, 150, 152, 156, 157
   born blue, 27
   breeding with, 151
   Celesky's Tulip, 63
   definition, **192**
   Francine, 112
   Ginger, 107, 109
   meat quality, 153
   unregistered, 146
   Ziggy, 107
Ginder, Tracy, 201t
Gladys, 72
goats, 81
   fainting, 79, 142
   milk, 125
   Pygmy, 110, 114, 115, 144
God's Blessing Farm, 139
Gonzales, Alesha, 91
GOS (Gloucestershire Old Spots), 22, 23, 24, 114, 138
government, US, 48, 49, 51, 64, 112
grazing, 11, 34, 35, 125
   and exercise, 198, 200
   on grass, 87
   Guinea Hogs good at, 41, 73, 105, 121
GRF (prefix), 201t
growth rate, 11, 54, 85, 94, 139, 146, 149, 177
Guinea Forest Hog. *see* Forest Guinea Hog
Guinea Hog Registry (GHR), 61, 62
Guinea Hogs, 22, 23, 140, 147, 150
   acceptable traits, 196
   AGHA purposes for, 65
   American Guinea Hog Selection Matrix, 150, 178
   American Guinea Hog Stewardship, 62
   appearance, 10, 11
   attributes, 139
   behaviors, 8–9, 11, 12
   being bred smaller, 115
   biodiversity for, 152
   black, 56, 104
   bloodlines, 11, 15, 180
   blue, 27, 44, 100, 101, 102, 122
   body types, 15, 41, 137
   breed description, 178, 197, 197–198, 199–200
   breed organizations, 1, 59–62
   breed standard, 32, 195, 196, 198, 200
   breeder prefixes, 201t
   breeding, 22, 81
   breeding stock, 138
   color, 16, 25, 25–27, 26, 30
   conservation, 24, 54, 56, 62, 65, 105, 122, 139, 154, 166, 175
   cost to rear, 12
   culinary qualities, 17–21
   decline, 56
   defining, 15, 30, 56
   diet, 11, 17
   disappearance, 48
   DNA, 3, 21–24, 22
   easy to handle, 121, 144
   education about, 65
   endangered lines, 133
   as fat supplier, 51
   as fatty heritage breed, 53
   flavor, 11, 12, 18
   gene pool, 203t
   genetic lines, 78, 128
   genetic testing, 168
   genetics, 16, 22, 23
   gentle nature, 92
   grazing, 41, 73, 105, 121
   growth of, 11
   handling, 12
   hardiness, 12, 23, 54, 149, 160, 171
   history, 12, 16–17, 43
   homestead characteristic, 144
   ideal traits, 90, 195
   importance in American culture, 10, 182
   inbreeding, 93
   increasing genetic diversity in, **194**
   as landrace breed, 14, 31, 64, 99, 199
   lard from, 199
   life expectancy, 143, 198, 200
   the Livestock Conservancy on, 17, 126, 179
   marketed as African Miniatures, 61
   meat, 7, 11, 17, 18, 34–35, 51–54, 91, 105, 158, 197
   monitoring, 60
   naming, 16, 17, 71
   needs, 12
   nominated for *Ark of Taste* catalog, 18

origins, 17, 21, 22, 23, 197
and pasture, 8, 53, 54, 98, 154, 167
personality, 7, 8, 9, 10, 165
pet pig craze impact on, 47
phenotype, 27
preservation, 65, 166, 181, 189
promoting, 56, 60, 64, 65, 150, 189
pure bloodlines of, 48
as rare breed, 121
recommended for breeding, 76
recovering, 152–157, 160–165, 167, 169
registering, 56, 59, 65
registry of, 4, 26, 27, 28, 65, 66, 105, 153
reintroduction of, 49
rooting, 79
size, 11, 15, 16, 59–60, 139
on small farms, 55
in Southeast United States, 10, 12, 16, 22, 23, 60, 104, 111, 112, 199
stamina, 141
stature, 61
tenacity, 154
tracking, 56, 64, 97, 143, 164, 177, 181
trainability, 11
Types A, B, and C, 61, 196
value, 44
weight, 15
Gulf Coast Native sheep, 39, 98, 99, 125, 150
Gullah, 12, 20

H

Haflinger horses, 110, 114, 116
hair, 26, 148. *see also* coat
  black, 26
  blue black, 88
  in breed description, 195, 196, 198, 200
  curly, 74, 91, 147
  density, 22, 23, 75, 80, 81, 89, 93, 115, 120, 141, 145, 147, 151
  length, 74, 91, 93
  red, 26
  straight, 74, 93
  white, 26, 31
Hale (line), 104–106, 133, 162, 172t, 180, 203t
Hale, Dan, 104–106, 107, 149, 172t, 203t
Hale, Shirley, 104–106, 107, 149, 172t, 203t

ham, 11, 12, 20, 21, 23, 43, 90
Ham Sweet Farm, 79
Hampshire (breed), 18, 34
hardiness, 12, 23, 54, 149, 160, 171
Harper, Rosa, 63, 181
head (body part), 80, 90, 148
  in breed description, 196, 197, 199
  shape of, 61, 75, 81
  use of, 12, 43
  white marks on, 29
headcheese, 12, 21
HeartsEase (prefix), 201t
herd books, 14, 66, 79, 133, 159, 160
herd dispersal, 91, 105, 109, 134, 146, 152, 153, 166
Hereford (breed), 22
Heritage (restaurant), 19
heritage breeds, 2, 11, 19, 116
  in breed description, 197, 199
  changes in, 115
  Dexter cattle are, 104
  endangered, 138
  fattiness of, 52, 53
  history of, 28
  hogs, 3, 53, 54, 56, 64, 140
  landrace types, 66
  livestock, 16, 54, 104, 106
  poultry, 16, 146, 150
  protecting, **191**
  rabbits, 125, 149
  raised by Cathy R. Payne, 150
  raised by Gabriella Nanci, 62
  raised by J. Frank Baylis, 79
  raised by the Hales, 104
  survival of, 34
heritage farms, 4, 19, 153
  in breed description, 197, 199
Hesters (Hale) Mork, 105, 107, 108, 172t, 203t
Hesters (line), 105, 114, 132, 134, 201t
  Arie McFarlen obtained stock from, 167
  bloodlines, 15, 98, 162
  characteristics, 147, 164
  entered into registry, 171
  genetics, 136, 148, 149
  managed by Becky Mahoney, 160, 163
  not entered in herd book, 133
  pedigrees, 168
  purchase of, 146
  Read and Hale lines of, 180

Hesters (prefix), 201t
  selection strategies, 148
Hesters, Annette, 97, 98, 105, 106–108, 109, 146, 147, 168, 201t
  on Hesters herd, 132, 149, 162, 170
Hesters Ajax, 148
Hesters Carlos, 105, 107
Hesters Charlie, 148
Hesters Cracker, 105, 107
Hesters Farm Log Homes, 106, 201t
Hesters Fiona, 148
Hesters Ginger, 107, 109, 172t, 203t
Hesters Iggy, 107, 162
Hesters Iggy II, 107
Hesters Molly, 107
Hesters Patti, 106, 107
Hesters Peter, 105, 107
Hesters Petunia, 105, 107
Hesters Polly, 106, 107
Hesters Ziggy, 107, 162
HHN (Historic Herds Network), 165–167, 168, 170, 171, 175
  breeders in, 201t
  code of ethics of, 165, 166
  definition, **192–193**
  evaluate hogs, **194**
  hogs of, 169
  organization of, 133–136
hips, 74, 77, 89, 145, 155
  in breed description, 199
hocks, 145, 155
  in breed description, 196
homestead hogs, 10, 22, 112, 144, 175, 182
homesteading, 49, 111, 148
  homesteaders, 10, 11, 146, 153
  homesteads, 3, 55, 56, 104, 150
horses, 14, 106, 120, 143, 144, 145, 155, **191**
House, Sandee, 103
Hunker, Matt, 71
Hurricane Katrina, 117, 121

I
Iberian pigs, 24
Idlewild Park, 111
Illinois, 67t, 84, 107, 109, 202t
Improved Essex (breed), 21–24
  breed description, 31
inbreeding, 93, 121, 143, 151, 157
  avoiding, **194**

coefficient, 93, 94, 98, 135, 140, 163, **191**, **193**
depression, 105, 177
for genetic similarity, 176
increases white marks, 25, 30, 31, 32, 72, 80, 116
linecrossing to avoid, 119, 142
reducing, **192**
Indiana, 88, 106, 131, 132, 146, 148
industrial farming, 47, 48, 49
Ingraham, Angela, 201t
Iowa, 64

J
Jacob sheep, 110, 115
Jail Creek Farm, 18
Jefferson, Thomas, 22
Jersey milk cows, 106
JNH (prefix), 201t
Jorgensen, Hans Peter, 1, 3, 7, 60
Joyful Noise Home-N-Stead, 146, 201t
Judy, 72

K
Kansas, 92
Keene, Fred, 27, 98–103, 101, 102
Keene, Suzanne, 102
Kentucky, 63, 86, 103, 143, 146, 169, 180
Khaki Campbell ducks, 150
King, Bob, 158
King, Karen, 158
King, Sam, 103
Krumm, Paul, 17, 62, 63, 64, 65, 84, 201t

L
Landner, Sebron, 99
landrace breeds, 39, 98, 179
  acceptable phenotypes of, 27
  characteristics acceptable in, 178
  conservation of, 134, 159
  definition, 13
  and genetic diversity, 15, **193**
  as genetic resources, 14
  Guinea Hogs as, 31, 64, 199
  and heritage breeds, 66
  inclusiveness in, 160
  and selection, 29, **194**
  standards for, 15, 28
lard, 7, 12, 21, 41, 49, 51
  in breed description, 197
  Guinea Hogs for, 199

production of, 22, 43, 126, 182, **194**
used for preservation, 42, 44
lard hogs, 24, 53, 56, 112, **191**
  definition, **193**
Large Black, 22, 138
Latin American breeds, 24
Laughing Stock Farm, 201t
legs, 93, 145, 155, 182
  in breed definition, 195
  in breed description, 196
  color on, 72
  length of, 75, 81, 88, 89, 101, 102, 157, **193**
  shape of, 77, 100
linebreeding, 74, 85, 90, 107, 136, 141, 160, 163, 167, 170, 175
  alternated with linecrossing, 149, 177
  definition, **193**
  and inbreeding depression, 105
  vs. linecrossing, 94
  to preserve lines, 135, 142, 162
  purpose of, 176
  and selecting for color, 31
linecrossing, 23, 135, 141, 162, 175
  in alternate generations, 94
  alternate linebreeding with, 149, 177
  to avoid excessive inbreeding, 125, 142
  definition, **193**
  drawbacks to, 176
  to infuse genetic diversity, **193**
Little Old Stiff Guy, 15, 67t, 85t, 86, 90, 91, 202t
Little-boned Boar, 131, 157, 158
little-boned variety, 8, 15, 23, 41, 56, 59, 105, 117, 119, 120, 131, 137, 147, 153, 157, 158. *see also* big-boned variety
livestock, rare, 2, 63
livestock breeds, 2, 39, 54, 104
Livestock Conservancy, The, 3, 98, 131, 133, 151, **191**, **193**. *see also* ALBC (American Livestock Breeds Conservancy); AMBC (American Minor Breeds Conservancy)
  advice on American Guinea Hog Selection Matrix, 179
  advice on herd placement, 108
  advice to breeders, 4
  assists conservation of rare breeds, 2
  conferences of, 148
  on disappearance of swine breeds, 47
  genetic analysis of pig breeds, 23

information on Guinea Hogs, 126
  involvement in, 150
  mission of, 189
  receives Sustainable Agriculture Research and Education (SARE) grant, 22
  researches Guinea Hogs, 17
local breeds, 11, 13, 14, 23, 159
Louise, 71, 72, 73, 74
LSF (prefix), 201t
LSF Bess, 74

M

Magby, Jessica, 110–116
Mahoney, Becky, 133, 134, 136, 137, 167, 171
  and bred gilts, 150
  as breeder, 201t
  in Genetic Recovery project, 108
  helps create HHN, 146–149
  maintains Hale line, 109
  manages Hesters herd, 132, 160, 163
  recovering the Guinea Hog, 152, 160, 164
  on selection goals, 178
  selection strategies, 106, 146, 148–149
Mahoney, Patrick, 146
Mahsrow (prefix), 201t
Maine, 72, 76
Maloy, Noah, 40
Managing Breeds for a Secure Future (Sponenberg et al.), 2, 13, 28, 53, 128, 136, 150, 175, 177
  book study group, 132, 134, 151
  guidelines of, 135, 152, 160, 165, 178
  strategies, 176
Mangalitsa (line), 29, 53
Maryland, 72, 73
Maveric (line), 153, 157, 161–165, 168, 169, 180
Maveric (prefix), 201t
Maveric Balthazar, 17
Maveric Big Boar, 127, 131, 137, 152, 155, 156, 165
Maveric Charles Sm 3, 122, 156, 163
Maveric Esther, 27, 103
Maveric Hansel, 119, 152, 155
Maveric Heritage Ranch, 17, 18, 27, 62, 98, 126, 162, 180, 201t
Maveric pedigrees, 156, 170

Maveric PeeWee, 103
Maveric Sadie, 127, 131, 137, 152, 157
MC (prefix), 85, 87
McFarlen, Arie, 17, 98, 100, 101, 106, 108, 117, 118, 122, 131, 133, 137, 156, 164, 167
   on blue in Guinea Hog, 27
   as breeder, 201t
   creates American Guinea Hog Stewardship, 62
   kept closed herd, 180
   nominates Guinea Hog for *Ark of Taste* catalog, 18
   preserves genetics, 106
   trades stock, 107
meat, 4, 43, 126
   breeding for, 180
   color, 21
   fat content, 20, 47, 48
   flavor, 11, 12, 17, 18, 44
   Guinea Hogs for, 7, 91, 197
   preservation of in lard, 42
   quality, 138
meat hogs, 127, 131, 136, 137
medical research, 16
Meishan (line), 53
Michigan, 79, 104, 107, 172t, 203t
Middleton Place Plantation, 12
Midwestern United States, 12, 18
Miniature African Pig, 16, 61
miniature donkeys, 110, 114
miniature Guinea Hogs, 16, 32, 56
Minnesota, 62, 63, 65, 140
Minnesota miniature hogs, 16
Minor Breeds Conservancy, **194**
   definition, **193**
Mississippi, 21, 27, 28, 35, 39, 97, 98, 100, 101, 116, 121, 122, 127, 128–131, 152–157, 158, 161, 166, 168, 169, 172t, 203t
Missouri, 64
Moffit, Lola, 60, 61, 195
Moore, Graeme (Gra'), 19, 20, 24, 26, 153, 201t
Morris Farm, 76, 78
mothering skills, 33, 141, 146, 149, 171, 200
   in breed description, 198
MS (prefix), 122, 166, 167
Mt. Citra Farm, 158
Mulefoot breed, 22, 146

N

Nanci, Gabriella, 60, 61, 62, 63, 112, 195
national herd, 78, 136, 143, 170, 178, **194**
natural breeds, 13
nature. *see* personality
Naumann, Lisa, 127
Nebraska, 62, 67t, 83, 97, 135, 202t
New Hampshire, 75, 76
New York, 92, 113
Nineteen61, 18, 157
North Carolina, 24
Northern Marsh Farm, 139
Northern United States, 12, 141, 148
nose, 10, 91, 104. *see also* snout
   breadth, 196
   digging with, 11
   long, 87, 122, 153, 157
   medium dished, 61, 196
   short, 74, 80, 141
   spot of pink on, 111
   white on, 26, 31, 61, 87, 198, 200
Nova Scotia, 79

O

Oberdorfer, Don, 31, 59, 62, 63, 64, 65, 72, 74, 88, 92, 97, 108, 133, 178, 179, 201t
   on established herds in Mississippi, 117
   helps form AGHA, 83
   on Hesters herd, 109
   president of AGHA, 35
   recommends Guinea Hogs, 76
   selection strategies, 32, 93
   on Setty stock, 87
OGS (prefix), 201t
Ohio, 63, 67t, 72, 85, 88, 90, 97, 135, 202t
Oklahoma, 146
Old Freedom Church, 201t
Old Freedom (prefix), 201t
Old Orchard Farm, 107, 110, 111
oral histories, 10, 14, 30, 133, 135, 170
Oregon, 17
Oregon Department of Agriculture, 17
Ossabaw Island (breed), 22, 24, 29, 53
outbreeding. *see* linecrossing 177

P

pancetta, 21
parasites, 33, 34, 180
pastern, 142, 145, 155

pasture, 20, 39, 48, 52, 76, 81, 87, 125, 180, 198
   flavor of hogs raised on, 12, 17, 18, 126, 138, 157
   grazing in, 11
   Guinea Hogs give birth in, 8
   raising Guinea Hogs on, 53, 54, 154, 167, 178, 200
   and supplemental feeding, 34
Payne, Cathy R., 72, 149–152, 179
   as breeder, 201t
   cofounds HHN, 165–167
   creates American Guinea Hog Selection Matrix, 178
   recovering the Guinea Hog, 136–137, 160, 161–165
   rehoming hogs, 140
   website of, 189
pedigrees, 28, 62, 65, 74, 79, 94, 106, 107, 122, 136, 140, 144, 156, 157, 161, 162, 164, 166, 167, 168, 169, 170, 177
   analysis, 22
   breeding prefixes in, 201t
   Facebook page on, 137
   gallery on website, 151, 189
   and inbreeding, 135
Pennsylvania, 107, 110, 115, 131, 172t, 203t
Perkins, Anna, 201t
Perkins, James, 201t
personality, 71, 73, 145, 154, 156, 165
   breed consistencies in, 13
   calm, 105, 171
   docile, 77, 112
   easy to manage, 10, 92, 144, 171
   easygoing, 63, 121, 148
   floppiness, 9, 91, 142, 148
   friendly, 144
   gentle, 91, 92, 101
   good natured, 48, 86, 89, 175
   independent, 143
   little, 89
   lots of, 7, 74, 182
   pleasant, 195
   unique, 151
pet pigs, 16, 22, 33, 47, 55–56, 59, 60, 105, 115
Petrini, Carlo, 18–19
phenotype, 10, 13, 24, 63, 104, 106, 160, 165, 171, 181, **192**
   definition, **193**

of Guinea Hogs, 27
piggies. *see* piglets
piglets, 8, 60, 66, 151, 156
   appearance of, 9
   and Celesky's Tulip, 85
   coloring of, 26, 27, 35, 61, 87
   definition, **194**
   durability of, 86
   human care of, 53
   litter size, 15, 90, 98, 122
   quality of, 142, 146, 160
   registering, 181
   selecting, 178
   sow's behavior toward, 10, 73, 74, 105, 198, 200
   temperament of, 148
Pine Tacky horses, 39, 98, 99
Pineyrooters, 40
Pineywoods cattle, 39, 98, 99
Pineywoods Guinea, 16
Pineywoods hogs, 99
Pleasant Ridge (prefix), 201t
policies, 49, 50, 54, 133, 134
population bottleneck. *see* genetic bottleneck
pork, 42, 139, 144
   bellies, 12, 20
   chops, 18
   fat content, 48, 49, 52, 54
   from Guinea Hogs, 17, 158
   home-raised, 140
   industry, 50
   most grown in CAFOs, 53
   pastured, 126, 138, 157
   production of, 54
   purchasing from breeders, 182
   rinds, 12
   in scrapple, **194**
   selling, 11
   ways served in restaurants, 20
pot-bellied pigs, 16, 31, 35, 56, 60, 61, 105, 114, 115
poultry, 2, 16, 125, 128, 189. *see also* chickens
Prichard, Diana, 52
Priest, James, 15–16, 39, 43
primitive breeds, 13
Punch, 72
purebred hogs, 16, 29, 30, 32, 33, 59, 80
   definition, **192**
Pygmy goats, 111, 114, 144

INDEX

R
rabbits, 43, 125, 149, 150
Rapp, Callene, 64, 76, 92, 110, 201t
rare breeds, 2, 3, 19, 22, 30, 54, 60, 66, 114
   protecting, 79
   pure Guinea Hogs are, 115
rare variants, 25, 26, 27, 28, 30
Read (line), 110, 133, 147, 180
Read, Marcia, 98, 107, 110–116, 131, 162, 172t, 203t
Read, Paul, 107, 110, 172t, 203t
   selection strategies, 115
Read Iggy, 107, 172t, 203t
Read Ziggy, 107, 172t, 203t
recessive genes, 26, 29, 35, 200
   in breed description, 198
Record of Improved Essex Swine, Volume III, 23
red bloodlines, 26, 162
red hogs, 22, 26, 29, 31, 80, 198, 200
Red Wattle, 22–23
Redmond, Tammy, 91
Registered American Guinea Hog Genetic Discussions, 128, 150
registration, 30, 84, 93, 94, 156, 167, 168, 189
   AGHA Guinea Hog registry, 4, 26, 27, 28, 65, 66, 105, 153
   formal registries, 4, 61, 109
   GHR (Guinea Hog Registry), 61, 62
   of Guinea Hogs, 4, 26, 27, 28, 65, 66, 105, 153
rendering, 42, 52, 126, **192**
   definition, **194**
resistance, 134
   to disease, 14, 22, 112, **193**
   to parasites, 33, 180
Roger Williams Park Zoo, 72, 73
rooting, 73, 79, 139, 141, 198, 200
Ross, Don Jr., 86, 91, 201t
Rowland, Jack, 92
Rudugast (prefix), 201t
Rudugast Farm, 201t
Rudy, 144, 154, 167
Russian hogs, 103

S
salt meat, 41
SARE (Sustainable Agriculture Research and Education) grant, 22
Sasha, 144, 154, 167
sausage, 21, 42, 43, 44, 140
Schulze, Ethan, 201t
Scottish Blackface sheep, 81
scrapple, 12, **194**
SCZ (prefix), 84, 201t
SCZ Boaris, 92
SCZ Bullwinkle NR002, 108
SCZ Fred, 92
SCZ Hampton, 92
SCZ Kat, 92
SCZ Kit, 92
Sebron Ladner Place, 99
Sedgwick County Zoo, 28, 64, 76, 84, 84–85, 108, 110, 201t
   Celesky hogs at, 92–93
selection, 114, 117, 178–181
   AGH Selection Matrix, 150, 178
   breeder role in, 14, 32
   can change gene frequencies, 175, 176
   characteristics for, 36, 81, 134, 148, 150, 154, 164
   for color, 30, 31, 33
   criteria for, 34, 80, 145, 178
   definition, **194**
   difficulty making, 29
   drivers of, 13, 28
   due to preference, 25
   for floppiness, 142
   for genetics, 180
   for hardiness, 160
   and inbreeding, 32, 93
   for looks, 179
   for more uniform traits, 177
   need for, 30
   for parasite resistance, 180
   process of, 156
   for short face, 80
   for size, 106, 108, 149
   strategies, 181
   to survive on pasture, 167
   within a closed herd, 180
Setty (line), 31, 97, 135, 139, 201t
   background, 25
   bloodlines of, 15
   carries white, 26, 31
   close relations in, 93–94
   DNA, 78
   not feral looking, 74
   rectangular shape, 92
   relative size, 141

Setty (prefix), 201t
Setty, Bonnie Jean, 85
Setty, Randy, 27, 63, 67t, 84, 90, 201t
    as foundation hog breeder, 172t, 202t, 203t
    herd of, 85–87
Setty Blackbart, 172t, 203t
Setty Blackjack Luther, 86, 91
Setty Houdini, 27, 31, 32, 63, 67t, 85, 88, 202t
    death of, 89
    personality of, 74
    as sire, 64, 87, 93, 94
Setty Lilly, 67t, 88, 202t
Setty Lucky, 172t, 203t
Setty Rose, 27, 31, 63, 64, 67t, 85, 87, 88–89, 94, 202t
Setty's MC Big Old Stiff Boar, 67t, 85, 90, 202t
Setty's MC Little Old Stiff Guy, 67t, 85, 90, 91, 202t
Setty's MC Wart Side Sow, 67t, 85, 90, 202t
sheep, 14, 60, 106
Shetland sheep, 63, 106
Shetland Sheepdogs, 106
Shirley (line), 118, 119, 153, 167, 170
Shirley, Dean, 156, 167
Shirley, Rhonda, 156, 167
shoats, 60, 84, 128, 130, 132, 139, 156, 170
    definition, **194**
shoulder (meat), 21, 43
shoulders (body part), 74, 77, 89, 90, 145, 155
    in breed description, 199
Silver Fox meat rabbits, 150
Silvera, Angela, 139
Silvera, Ricardo, 139
Silvera, Rico, 117, 162
Sinclair hogs, 16
sire, 75, 107, 168, 169, 170, **193**
    definition, **194**
skin, 114, 169, **192**
    in breed description, 195, 196
    coloring, 10, 26, 27
Skyfire, 17
Skyfire (prefix), 201t
Skyfire Farm, 17
Skyfire Garden Seeds, 201t
Skyfire Heirloom, 17
slaughtering. *see* butchering

Slow Food Charleston, 19, 20, 24
Slow Food International, 12, 18
Slow Food USA, 18, 19
Small Black (breed), 23
small breeds, 54, 86, 177, 179
small farms, 50, 55, 151
small landholders, 3, 10, 12, 126, 132
small-boned variety. *see* little-boned variety
Smith, Carmen, 201t
Smith, Frances, 201t
Smithfield Foods, 53
smokehouses, 21, 42, 43, 44
snout, 87, 92, 141, 145, 155. *see also* nose
    in breed description, 197, 199
    coloring of, 25, 26, 198, 200
    definition, 196
socks, 33, 87, **191**. *see also* feet (body part)
Solomon's Wisdom Farm, 103, 143, 167, 180, 201t
souse, 12, 43
South Carolina, 8, 12, 19, 20, 40, 100
South Dakota, 63, 99, 100, 107
Southdown sheep, 116
Southeast United States, 42, 79, 111, 141
    butchering time in, 1
    chitterlings in, **191**
    farming changes in, 48
    Guinea Hog in, 10, 12, 16, 22, 23, 60, 104, 199
    hog as part of cultural history, 3, 7
    livestock roamed, 99
    people of, 7, 12
    weather in, 10, 113
sows, 132, **192**
    blue, 26, 44
    definition, **194**
    manageability, 86
    mothering skills, 141
    number of litters, 31
    size, 15
Spencer, Gary, 59, 60, 61
Spinillo, Christian, 79
Spinillo, Kate, 79
Sponenberg, D. Phillip, 13, 14, 23–24, 25, 28, 29, 30, 97, 100, 125, 128, 133, 134, 149, 159, 163, 175, 176, 177, 179
Sprague Maveric Buster, 103
Stacy, 113, 114
stance, 145
standardized breeds, 12, 13, 27, 28, 176

Stanfield, Jocelyn, 91
strains, 1, 4, 15, 26, 61, 78, 115, 136
   as breeding consideration, 36
   characteristics of, 126
   commercial, 14
   family, 127
   foundation, 135, 167
   industrial, 50–54, 197, 199
   isolated, 134
   mixed, 16
   preserving, 160
   rare, 133
Sugars (prefix), 201t
Sugars Shadey Adie, 87, 91
Sullbar (prefix), 201t
Sullbar Farm, 32, 76, 78, 79, 88, 135, 201t
Sullbar VA ML Porgy, 75, 79
Sullivan, Shirley, 32, 35, 75, 76, 88, 133, 201t
Sumrall (line), 15, 76, 97, 119, 133, 136, 160, 162
   characteristics of, 164
   dispersal of, 152
   in Donna Dorminey's herd, 167
   entered into AGHA registry, 171
   genetics, 151
   pedigree, 161
   perpetuation of, 137, 153, 154, 163
   a rediscovered old herd, 141
   relations between, 157
Sumrall (prefix), 201t
Sumrall, Brent (Buck), 117, 118, 127–128, 130, 131, 136, 137, 140, 152, 153, 154, 155, 156, 158, 161, 164
Sumrall, Gary, 21, 98, 116, 117, 119, 122, 128, 131, 133, 140, 154, 162, 201t, 203t
   breeding standards of, 157
   as foundation hog breeder, 172t
   linecrossing, 119
   selection goals of, 178
   selection strategies, 35, 130, 155, 156
   as selective breeder, 156
Sumrall, Harmon David, 116
Sumrall, Janis, 130, 136, 152, 154, 156
Sumrall, John William, 116, 120
Sumrall Bobbie Sue, 15, 122, 137, 156, 162, 163, 166, 167
   as foundation sow, 172t, 203t
Sumrall family, 116–122, 117, 130, 131, 137, 153, 168
   feeder hogs of, 157–158

Sumrall Farm, 153, 160, 170, 201t
Sumrall Mississippi Belle, 169
Sumrall MS Big Momma, 131, 141, 152
Sumrall MS DuBose, 169, 170
Sumrall MS Swanee Rose, 160, 170
Sunrise (prefix), 201t
Sunrise Farm, 201t
Sustainable Agriculture Research and Education (SARE), 22
sustainable farming, 3, 198, 200
Sustainable Farming magazine, 178
sustainable heritage farm pig, 197
SWF (prefix), 201t
SWF MS Breandon, 103
SWF MS Lenah, 103
SWF MS Willie Mae, 103
swine, 10, 11, 168, **193**, **194**
   breeds, 3, 13, 22, 47
   industrial strain of, 53
   landrace, 13, 14
   treatment of, 166

T
tail, 10, 75, 81, 141, 147
   in breed definition, 195
   in breed description, 196, 197, 199, 200
   docking, 41, 53
Tamworth, 23, 114
Taylor, Erin M. (Micki), 62, 64
teats, 141, 151, 171
temperament. *see* personality
Tennessee, 161
thick-boned variety. *see* big-boned variety
Thomas, Shan, 60
thriftiness, 98, 134, 171, 178, 182, **192**
   definition, **194**
toes, 87, 92, 145
   in breed definition, 195
   in breed description, 196
tracking Guinea Hogs, 56, 64, 97, 143, 164, 177, 181
traits, 30, 31, 33, 34, 176, 181
   adapted, 14, 134, 159
   biological, 14
   in breed description, 195, 196
   breed standard, 15
   breeding for, 179
   color, 28, 29, 32

continuation of, 142
for curly hair, 74
genetic, 13, 28
landrace, **193**
natural, 54
rare, 29
selecting for, 33, 155
Sumrall, 154
survival, 180
unacceptable, 61
uniform, 177
Tunis sheep, 110, 115
tusks, 71, 73, 89, 157

U
UC Davis Veterinary Genetics Laboratory, 167, 168, 169
   definition, **194**
Udder Hope Matt, 77
UGF BB King, 91
USDA (US Department of Agriculture), 52, 54, 113

V
VA (prefix), 75, 78, 79, 135
vaccination, 34
Vala's Pumpkin Patch, 83
variant, 25, 27, 28, 30, **191**
variations, 14, 26, 27, 28
   in breed description, 199
   color, 25, 29, 198, 200
VAZ (prefix), 201t
VAZ Harriet, 74
VAZ John Henry, 74
Vietnamese pot-bellied pigs, 16, 55–56, 60–61, 105, 114, 115
vigor, 94, 135, 151, 176, 177, 178, 182, **194**
Virginia, 62, 67t, 73, 75, 76, 79, 81, 86, 97, 107, 109, 172t, 202t, 203t
Virginia Zoo, 201t

W
Watershed Restaurant, 18
Watkins (line), 84, 108, 133, 162
   Celesky hogs have a trace of, 115
Watkins, Donna, 67t, 84, 98, 107, 108–109, 202t
   stock of, 27
weaknesses, 141, 142, 160, 176
   in breed description, 196
weight, 145, 155
   of adult Guinea Hog, 15
   in breed definition, 195
   of carcass, 11, 20, 105, 200
   excess, 88, 89, 113
   finished, 54
   gained easily, 8, 85
   recorded, 92
WH Group of China, 53
white feet, 25, 26, 29, 30, 31, 32, 33, 34, 35, 80, 87, 111
   in breed description, 196
wild breeds, 13, 74, 80, 100, 171
Wilma, 104
Wisconsin, 76, 138, 153, 160

Y
Yard Pig, 16
Yearington, Miranda, 201t
Yelverton, Brian, 201t
Yokeley's Summer Thyme, 8

www.ingramcontent.com/pod-product-compliance
Lightning Source LLC
Chambersburg PA
CBHW052021070526
44584CB00016B/1850